SpringerBriefs in Mathematics

SpringerBriefs in Mathematics showcases expositions in all areas of mathematics and applied mathematics. Manuscripts presenting new results or a single new result in a classical field, new field, or an emerging topic, applications, or bridges between new results and already published works, are encouraged. The series is intended for mathematicians and applied mathematicians.

BCAM SpringerBriefs

BCAM *SpringerBriefs* aims to publish contributions in the following disciplines: Applied Mathematics, Finance, Statistics and Computer Science. BCAM has appointed an Editorial Board, who evaluate and review proposals.

Typical topics include: a timely report of state-of-the-art analytical techniques, bridge between new research results published in journal articles and a contextual literature review, a snapshot of a hot or emerging topic, a presentation of core concepts that students must understand in order to make independent contributions.

Please submit your proposal to the Editorial Board or to Francesca Bonadei, Executive Editor Mathematics, Statistics, and Engineering: francesca.bonadei@springer.com

More information about this series at http://www.springer.com/series/10030

Gengsheng Wang · Yashan Xu

Periodic Feedback Stabilization for Linear Periodic Evolution Equations

Gengsheng Wang
School of Mathematics and Statistics
Wuhan University
Wuhan
China

Yashan Xu
School of Mathematical Sciences
Fudan University
Shanghai
China

ISSN 2191-8198 ISSN 2191-8201 (electronic)
SpringerBriefs in Mathematics
ISBN 978-3-319-49237-7 ISBN 978-3-319-49238-4 (eBook)
DOI 10.1007/978-3-319-49238-4

Library of Congress Control Number: 2016960559

Printed on acid-free paper

This Springer imprint is published by Springer Nature
The registered company is Springer International Publishing AG
The registered company address is: Gewerbestrasse 11, 6330 Cham, Switzerland

Preface

Stability theory was first established by Aleksandr Lyapunov in 1892 (see [70]). Due to the wide application of stability theory, many mathematicians are devoted to this area. For an unstable differential system, how to stabilize it, with the aid of a feedback control, becomes an important subject in control theory of differential equations. This is termed feedback stabilization. Studies on this subject started with finite dimensional systems in the 1950s and extended to infinite dimensional systems in the 1960s. This subject contains two important themes: criteria to judge whether a controlled system is feedback stabilizable and design of feedback laws to stabilize systems.

In most past publications on criteria of stabilization, control systems have been linear and time invariant. There are quite limited studies on criteria of the periodic stabilization for linear and time-periodic controlled systems. The reason for studying the latter can be explained as follows: Mature theories have been established on the stability and stabilization for time-invariant linear ODEs. Regarding stability, it is a well-known result that the equation: $\dot{y}(t) = Ay(t), t \geq 0$, with $A \in \mathbb{R}^{n\times n}$, is exponentially stable if and only if the spectrum of A is in the half plane $\mathbb{C}^- \triangleq \{z \in \mathbb{C} : Re(z) < 0\}$(see, for instance, [2]). The most important result regarding stabilization is Kalman's criterion: A pair of matrices $[A, B]$ in $\mathbb{R}^{n\times n} \times \mathbb{R}^{n\times m}$ is stabilizable (i.e., there exists a matrix K in $\mathbb{R}^{m\times n}$ so that the spectrum of $(A + BK)$ is in $\mathbb{C}^-$) if and only if the rank of $(\lambda I - A, B)$ equals to n for each $\lambda \in \mathbb{C}\backslash\mathbb{C}^-$ (see, for instance, [24, 86]). With respect to the stability of linear time-periodic ODEs, one of the most important results is that the periodic equation: $\dot{y}(t) = A(t)y(t), t \geq 0$, with $A(\cdot)$ T-periodic in $L^\infty(\mathbb{R}^+; \mathbb{R}^{n\times n})$, is exponentially stable if and only if the spectrum of $\Phi_A(T)$ belongs to the open unit ball in $\mathbb{C}$, where $\Phi_A(\cdot)$ is the fundamental solution associated with $A(\cdot)$ (see, for instance, [38, 40]). In view of Kalman's criterion on the stabilization of time-invariant pairs and the above-mentioned criterion on the stability of periodic equations, it is natural to ask for criteria on the periodic stabilization for periodic pairs.

When a pair $[A,B] \in \mathbb{R}^{n\times n} \times \mathbb{R}^{n\times m}$ is stabilizable, any matrix $K \in \mathbb{R}^{m\times n}$, with the spectrum of $(A+BK)$ in $\mathbb{C}^-$, is called a feedback stabilization law for the pair $[A,B]$. It is important to find ways to construct feedback laws. The usual structure of feedback laws is connected with either the LQ theory, as well as Riccati equations, or Lyapunov functions (see, for instance, either Chap. 9, [59] or Chap. 5, [86]). Correspondingly, how to construct periodic feedback stabilization laws for a given stabilizable T-periodic pair should also be important.

According to our understanding, for each unstable T-periodic $A(\cdot) \in L^\infty(\mathbb{R}^+;\mathbb{R}^{n\times n})$, the procedure to stabilize periodically the system: $\dot{y}(t) = A(t)y(t), t \geq 0$, is as follows: First, one builds up a T-periodic $B(\cdot) \in L^\infty(\mathbb{R}^+;\mathbb{R}^{n\times m})$ so that $[A(\cdot),B(\cdot)]$ is T-periodically stablizable; Second, one designs a T-periodic $K(\cdot) \in L^\infty(\mathbb{R}^+;\mathbb{R}^{m\times n})$ so that $A(\cdot)+B(\cdot)K(\cdot)$ is exponentially stable. We call the aforementioned $B(\cdot)$ a control machine and the corresponding $K(\cdot)$ a feedback law. It should be interesting to study the question of how to design a *simple* T-periodic $B(\cdot)$ for a given T-periodic $A(\cdot)$ so that $[A(\cdot),B(\cdot)]$ is T-periodically stabilizable. Of course, we can define what *simple* means according to our needs.

The aim of this monograph is to present recent advances regarding periodic stabilization for some linear and time-periodic evolution equations which contain both finite and infinite dimensional systems. These advances may lead us to a comprehensive understanding of the subject of periodic stabilization. The monograph summarizes our ideas, results, and methods with respect to the subject during recent years. Insofar as possible, we have tried to make the material self-contained. There is much literature on the stabilization of differential equations, and we are unable to give a complete list of references. Consequently, it is possible that some important works in the field will have been overlooked.

The monograph is organized as follows: Chapter 1 presents some preliminaries on linear periodic evolution equations, in particular, the connection between the LQ theory and periodic stabilization. Chapter 2 studies the periodic stabilization for some infinite dimensional linear periodic evolution equations. Three criteria on the periodic stabilization for a linear periodic evolution equation are provided. One is a geometric condition which is related to the attainable subspaces, while the other two are analytic conditions which are connected with some unique continuation properties of dual equations. Some applications of these criteria are also given in this chapter. Chapter 3 provides two criteria on the periodic stabilization of periodic linear ODEs. One is an algebraic condition which is an extension of Kalman's crite- rion to the periodic case, while the other is a geometric condition which is connected with the null-controllable subspace. Chapter 4 shows how to find a *simple* control machine $B(\cdot) \in L^\infty(\mathbb{R}^+;\mathbb{R}^{n\times m})$, for a given unstable periodic $A(\cdot) \in L^\infty(\mathbb{R}^+;\mathbb{R}^{n\times n})$, so that $[A(\cdot),B(\cdot)]$ is T-periodically stablizable. This is an application of the geometric criterion in Chapter 3.

The authors would like to acknowledge support from the Natural Science Foundation of China under grants 11571264, 11471080, 11171264, and 91130022.

The authors thank Professor Emmanuel Trélat for introducing us to several interesting papers relating to the subjects covered in our monograph. The authors would like to express their great appreciation to Professor X. Zhang and Professor E. Zuazua for the encouragement that they provided.

Wuhan, China
Shanghai, China
September 2016

Gengsheng Wang
Yashan Xu

Contents

Acronyms

Let X and Y are two real Hilbert spaces (identified with its dual). The following notations are effective throughout this monograph

$\|\cdot\|_X$	The norm of X
$\langle\cdot,\cdot\rangle_X$	The inner product of X
$\mathscr{L}(X,Y)$	The space of all linear bounded operators from X to Y
L^*	The adjoint operator of $L \in \mathscr{L}(X,Y)$
$\mathscr{N}(L)$	The kernel of $L \in \mathscr{L}(X,Y)$
$\mathscr{R}(L)$	The range of $L \in L(X,Y)$
$L(X)$	The space of all linear bounded operators from X to X
I_X	The identity operator on X (sometimes we simply write it as I)
$\sigma(L)$	The spectrum of $L \in \mathscr{L}(X)$
$L^\dagger$	The pseudo inverse of $L \in \mathscr{L}(X)$
$X \oplus Y$	The direct sum of X and Y
$Y^\perp$	The orthogonal complement of Y in X, when $Y \subset X$
X^C	The complexification of X, i.e., $x \in X^C$ if and only if there is x_1 and x_2 in X so that $(x_1,x_2) \in X \otimes_c X$, thereafter we write $x \triangleq x_1 + ix_2$ (where i is the imaginary unit)
L^C	The complexification of $L \in \mathscr{L}(X,Y)$, i.e., $L^C(x_1+ix_2) \triangleq Lx_1 + iLx_2$ for any $x_1,x_2 \in X$
$span\{x_1,\ldots,x_k\}$	The subspace generated by $x_1,\ldots,x_k$, when $x_1,\ldots,x_k \in X$
$\mathrm{Proj}_Y x$	The orthogonal projection of $x \in X$ into Y, when Y is a subspace of X
$\dim,\overline{Y}$	The dimension of Y, when Y is a finite dimension vector space
$Rank(Q)$	The rank of matrix Q
$\mathbb{B}(0,\delta)$	The open ball in $\mathbb{C}^1$, centered at the origin and of radius $\delta > 0$, and write $\mathbb{B} \triangleq \mathbb{B}(0,1)$
$\partial\mathbb{B}(0,\delta)$	The boundary of $\mathbb{B}(0,\delta)$
$\mathbb{R}^+$	The interval $[0,\infty)$

$L^p(E;X)$	The space of all Lebesque p-integrable functions on E with the value in X
$C(E,X)$	The space of all continuous functions on E with the value in X

Chapter 1
Controlled Periodic Equations, LQ Problems and Periodic Stabilization

Abstract In this chapter, we present some concepts and results related to the periodic feedback stabilization and LQ problems for some linear time-periodic evolution systems.

Keywords Periodic Equations · LQ Problems · Stabilization

1.1 Controlled Periodic Evolution Equations

The controlled evolution equation studied in this monograph is formulated as follows:

$$y'(s) = A(s)y(s) + B(s)u(s), \quad s \in \mathbb{R}^+ \triangleq [0, \infty). \tag{1.1}$$

Here and throughout this monograph, we suppose that the following main assumptions hold:

Main Assumptions:

($\mathscr{H}_1$) $A(s) = A + D(s)$, $s \geq 0$, where the operator A, with its domain $\mathscr{D}(A)$, generates a C_0 compact semigroup $\{S(t)\}_{t\geq 0}$ in a real Hilbert space H with the norm $\|\cdot\|$ and the inner product $\langle\cdot,\cdot\rangle$; $D(\cdot) \in L^1_{loc}(\mathbb{R}^+; \mathscr{L}(H))$ is T-periodic in time, i.e., $D(t+T) = D(t)$ for a.e. $t \in \mathbb{R}^+$. (Here and thereafter, $T > 0$ and $\|\cdot\|$ will be used to denote the usual norm of $\mathscr{L}(H)$ when there is no risk to cause any confusion.)

($\mathscr{H}_2$) $B(\cdot) \in L^\infty(\mathbb{R}^+; \mathscr{L}(U, H))$ is T-periodic. Here U is another real Hilbert space and controls u are taken from the space $L^2(\mathbb{R}^+; U)$.

For each $t \geq 0$, $h \in H$ and $u \in L^2(t, \infty; U)$, we also consider the following controlled equation:

$$y'(s) = A(s)y(s) + B(s)u(s), \quad s \geq t, \quad y(t) = h. \tag{1.2}$$

G. Wang and Y. Xu, *Periodic Feedback Stabilization for Linear Periodic Evolution Equations*, SpringerBriefs in Mathematics,
DOI 10.1007/978-3-319-49238-4_1

Definition 1.1 A mild solution of the Eq. (1.2) is a function $y \in C([t, \infty); H)$ verifying

$$y(s) = S(s-t)y_0 + \int_t^s S(s-r)\big[D(r)y(r) + B(r)u(r)\big]\mathrm{d}r \quad \text{for each } s \geq t. \tag{1.3}$$

Definition 1.2 Let X and Y be two Banach spaces. Write

$$E \triangleq \big\{(s,t) \in \mathbb{R}^+ \times \mathbb{R}^+ \;\big|\; 0 \leq t \leq s < +\infty\big\}. \tag{1.4}$$

(i) A function $F : \mathbb{R}^+ \to \mathscr{L}(X,Y)$ is strongly continuous at $s_0 \in \mathbb{R}^+$ if for any $x \in X$, $\varepsilon > 0$, there is $\delta \triangleq \delta(\varepsilon, s_0, x) > 0$ so that $\|F(s)x - F(s_0)x\|_Y < \varepsilon$, as $|s - s_0| \leq \delta$. If the above F is strongly continuous at each $s \geq 0$, then it is said to be strongly continuous over $\mathbb{R}^+$.

(ii) A function $\Phi(\cdot,\cdot) : E \to \mathscr{L}(X,Y)$ is strongly continuous at $(s_0,t_0) \in E$ if for any $x \in X$ and $\varepsilon > 0$, there is $\delta \triangleq \delta(\varepsilon, s_0, t_0, x) > 0$ so that $\|\Phi(s,t)x - \Phi(s_0,t_0)x\|_Y < \varepsilon$, when $(s,t) \in E$ and $|s - s_0| + |t - t_0| \leq \delta$. If $\Phi(\cdot,\cdot)$ is strongly continuous at each (s,t) in E, then it is said to be strongly continuous over E.

Definition 1.3 (*Evolution*) A function $\Phi(\cdot,\cdot) : E \to \mathscr{L}(H)$ is called an evolution generated by $A(\cdot) \triangleq A + D(\cdot)$ if it is strongly continuous and satisfies that

$$\begin{cases} \Phi(s,s) = I, & \text{when } s \in \mathbb{R}^+ \\ \Phi(s,r)\Phi(r,t) = \Phi(s,t), & \text{when } 0 \leq t \leq r \leq s, \end{cases} \tag{1.5}$$

and that when $0 \leq t \leq s$ and $h \in H$,

$$\begin{aligned} \Phi(s,t)h &= S(s-t)h + \int_t^s S(s-r)D(r)\Phi(r,t)h\mathrm{d}r \\ &= S(s-t)h + \int_t^s \Phi(s,r)D(r)S(r-t)h\mathrm{d}r. \end{aligned} \tag{1.6}$$

We will write Φ for $\Phi(\cdot,\cdot)$ if it will not cause any confusion. We would like to mention what follows: In Proposition 1.2, we prove that $A(\cdot)$ generates a unique evolution Φ over E. Hence, the commutativity property in (1.6) holds. The detailed proof of this commutativity property can be found in the proof of (i) of Proposition 1.2.

Proposition 1.1 *For each $t \geq 0$, $h \in H$ and $u \in L^2(t, \infty; U)$, the Eq. (1.2) has a unique mild solution, denoted by $y(\cdot; t, h, u)$.*

Proof The existence follows from the Picard iteration, while the uniqueness follows from the standard argument. This ends the proof. □

Proposition 1.2 *(i) The operator $A(\cdot) \triangleq A + D(\cdot)$ generates a unique evolution Φ over E. (ii) The evolution Φ generated by $A(\cdot)$ is T-periodic, i.e.,*

$$\Phi(s+T, t+T) = \Phi(s,t) \quad \textit{for all } 0 \le t \le s < \infty. \tag{1.7}$$

(iii) The mild solution $y(\cdot; t, h, u)$ to the Eq. (1.2), with $t \ge 0$, $h \in H$ and $u \in L^2(t, \infty; U)$, verifies that

$$y(s; t, h, u) = \Phi(s,t)h + \int_t^s \Phi(s,r)B(r)u(r)\mathrm{d}r, \quad s \in [t, +\infty). \tag{1.8}$$

Proof We beign with showing (i). By Proposition 1.1, for each $t \in \mathbb{R}^+$ and $h \in H$, Eq. (1.2), with the null control, has a unique solution $y(\cdot; t, h, 0)$ which verifies that

$$y(s; t, h, 0) = S(s-t)h + \int_t^s S(s-r)D(r)y(r; t, h, 0)\mathrm{d}r \quad \text{for any } s \ge t. \tag{1.9}$$

Define

$$\Phi(s,t)h \triangleq y(s; t, h, 0) \quad \text{for all } (s,t) \in E,\ h \in H. \tag{1.10}$$

It is clear that $\Phi : E \mapsto \mathscr{L}(H)$ is strongly continuous and satisfies both (1.5) and the first equality in (1.6). To show that the above defined Φ verifies the second equality in (1.6), we let $t \in \mathbb{R}^+$ and $h \in H$. Define

$$z(s) = S(s-t)h + \int_t^s \Phi(s,r)D(r)S(r-t)h\mathrm{d}r, \quad s \in [t, \infty). \tag{1.11}$$

By (1.9)–(1.11), using Fubini's theorem, one can easily check that $z(s) = y(s; t, h, 0)$ for all $s \ge t$. Then from the uniqueness of the mild solution, the second equality in (1.6) follows at once. To prove the uniqueness of the evolution, we let Ψ be another evolution satisfying (1.5) and (1.6). Given $t \ge 0$ and $h \in H$, set $y(s) \triangleq \Psi(s,t)h$ for all $s \ge t$. By the properties of Ψ, we find that $y(s) = y(s; t, h, 0)$ for all $s \ge t$. This, along with (1.10), indicates that Φ and Ψ are the same. Hence, (i) holds. (ii) follows from the T-periodicity of $D(\cdot)$ and (1.6), while (iii) follows from (1.3), (1.6) and Fubini's theorem. This ends the proof. □

The periodic map (or the Poincaré map) plays an important role in the studies of periodic systems. We next introduce its definition and some related properties.

Definition 1.4 Let Φ be the evolution generated by $A(\cdot)$. Write

$$\mathscr{P}(t) \triangleq \Phi(t+T, t), \ t \ge 0. \tag{1.12}$$

The map $\mathscr{P}(0)$ is called the periodic map associated with $A(\cdot)$.

Proposition 1.3 *Let $\mathscr{P}(\cdot)$ be given by (1.12).*

(i) For each $t \in \mathbb{R}^+$, $\mathscr{P}(t)$ is compact and the set $\sigma(\mathscr{P}(t)^C) \setminus \{0\}$ consists entirely of all nonzero eigenvalues of $\mathscr{P}(t)^C$.
(ii) The set $\sigma(\mathscr{P}(t)^C) \setminus \{0\}$ is independent of $t \in \mathbb{R}^+$.

(iii) If η is an eigenfunction of $\mathscr{P}(t)^C$ corresponding to an eigenvalue λ, then $\Phi(s,t)^C\eta$, with $s \geq t$, is an eigenfunction of $\mathscr{P}(s)^C$ corresponding to λ.

Proof By the compactness of $S(t)$ with $t > 0$, the assumption $(\mathscr{H}_1)$ and (1.6), one can easily check that for each $t \geq 0$, $\mathscr{P}(t)$ is compact. Consequently, each $\mathscr{P}(t)^C : H^C \to H^C$ is also compact. Then by the F. Riesz theorem, for each $t \geq 0$, $\sigma(\mathscr{P}(t)^C) \setminus \{0\}$ consists of all nonzero eigenvalues of $\mathscr{P}(t)^C$. Arbitrarily fix s_1 and s_2 with $0 \leq s_1 \leq s_2 \leq T$. Let

$$\mathscr{P}(s_1)^C\eta = \lambda\eta \text{ with } \lambda \in \mathbb{C}^1, \ \lambda \neq 0, \ \eta \in H^C. \tag{1.13}$$

Write $\lambda = \alpha_1 + i\alpha_2$ with $\alpha_1, a_2 \in \mathbb{R}$ and $\eta = \eta_1 + i\eta_2$ with $\eta_1, \eta_2 \in H$. By (1.13), we have

$$\mathscr{P}(s_1)\eta_1 = \alpha_1\eta_1 - \alpha_2\eta_2, \qquad \mathscr{P}(s_1)\eta_2 = \alpha_2\eta_1 + \alpha_1\eta_2. \tag{1.14}$$

From (1.14) and (1.7), one can easily check that

$$\mathscr{P}(s_2)^C(\Phi(s_2,s_1)^C\eta) = \lambda\Phi(s_2,s_1)^C\eta.$$

This implies that λ is also an eigenvalue of $\mathscr{P}(s_2)^C$ and $\Phi(s_2,s_1)^C\eta$ is a corresponding eigenfunction. Hence, we have that

$$\sigma(\mathscr{P}(s_1)^C) \setminus \{0\} \subseteq \sigma(\mathscr{P}(s_2)^C) \setminus \{0\}.$$

Similarly, by the T-periodicity of $\mathscr{P}(\cdot)$, we can show that

$$\sigma(\mathscr{P}(s_2)^C) \setminus \{0\} \subseteq \sigma(\mathscr{P}(s_1+T)^C) \setminus \{0\} = \sigma(\mathscr{P}(s_1)^C) \setminus \{0\}.$$

This completes the proof. □

Let

$$\hat{n} \triangleq \text{the number of all distinct eigenvalues } \lambda \text{ of } \mathscr{P}(0)^C \text{ with } |\lambda| \geq 1 \tag{1.15}$$

and

$$\hat{\delta} \triangleq \begin{cases} \max\left\{|\lambda| \mid \lambda \in \sigma\left(\mathscr{P}(0)^C\right) \setminus \{0\}, \ |\lambda| < 1\right\}, & \text{if } \sigma\left(\mathscr{P}(0)^C\right) \setminus \{0\} \neq \varnothing, \\ \frac{1}{2}, & \text{if } \sigma\left(\mathscr{P}(0)^C\right) \setminus \{0\} = \varnothing. \end{cases} \tag{1.16}$$

From the compactness of $\mathscr{P}(0)^C$, Proposition 1.3 and F. Riesz theorem, we find that $\infty > \hat{n} \geq 0$ and $\hat{\delta} < 1$. Write $\lambda_1, \ldots, \lambda_{\hat{n}}$ for all distinct eigenvalues of $\mathscr{P}(0)^C$, with $|\lambda_j| \geq 1$. Denote by l_j the algebraic multiplicity of λ_j. From Theorem 6.26 in [46], we have $\lambda_j < \infty$. Let

$$n_0 \triangleq l_1 + \cdots + l_{\hat{n}}. \tag{1.17}$$

We next introduce the Kato projection (see [46]), which plays another role in our studies.

Definition 1.5 Arbitrarily fix a $\delta \in (\hat{\delta}, 1)$, where $\hat{\delta}$ is given by (1.16). Let Γ be the circle $\partial\mathbb{B}(0, \delta)$ with the anticlockwise direction in $\mathbb{C}^1$. For each $t \geq 0$, the following operator is called a Kato projection associated with $\mathscr{P}(t)$:

$$\mathscr{K}(t) \triangleq \frac{1}{2\pi i}\int_{\Gamma} (\lambda I - \mathscr{P}(t)^C)^{-1} \mathrm{d}\lambda. \tag{1.18}$$

For each $t \geq 0$, let $\big(I - \mathscr{K}(t)\big)\big|_H$ be the restriction of $\big(I - \mathscr{K}(t)\big)$ on H. Write

$$\mathbb{P}(t) \triangleq \big(I - \mathscr{K}(t)\big)\big|_H. \tag{1.19}$$

Proposition 1.4 *Let $\mathscr{P}(\cdot)$ and $\mathbb{P}(\cdot)$ be defined by (1.12) and (1.19) respectively. Then each $\mathbb{P}(t)$ (with $t \geq 0$) is a projection on H so that*

$$H = H_1(t) \bigoplus H_2(t), \tag{1.20}$$

where

$$H_1(t) \triangleq \mathbb{P}(t)H \ \ \text{and} \ \ H_2(t) \triangleq (I - \mathbb{P}(t))H. \tag{1.21}$$

Moreover, $\mathbb{P}(\cdot)$, $H_1(\cdot)$ and $H_2(\cdot)$ have the following properties: (a) $\mathbb{P}(\cdot)$, $H_1(\cdot)$ and $H_2(\cdot)$ are T-periodic; (b) For each $t \geq 0$, both $H_1(t)$ and $H_2(t)$ are invariant subspaces of $\mathscr{P}(t)$; (c) If $\hat{n}$ and n_0 are given by (1.15) and (1.17), then

$$\sigma(\mathscr{P}(t)^C\big|_{H_1(t)^C}) = \{\lambda_j\}_{j=1}^{\hat{n}}, \qquad \dim H_1(t) = n_0; \tag{1.22}$$

(d) When $0 \leq t \leq s < \infty$, $\Phi(s, t) \in \mathscr{L}(H_j(t), H_j(s))$ with $j = 1, 2$; (e) When $0 \leq t \leq s < \infty$,

$$\Phi(s, t)\mathbb{P}(t) = \mathbb{P}(s)\Phi(s, t); \tag{1.23}$$

(f) Let $\hat{\rho} \triangleq (-\ln \hat{\delta})/T > 0$ with $\hat{\delta}$ given by (1.16). For any $\rho \in (0, \hat{\rho})$, there is $C_\rho > 0$ so that

$$\|\Phi(s, t)h_2\| \leq C_\rho e^{-\rho(s-t)}\|h_2\|, \ \text{when} \ \ 0 \leq t \leq s < \infty \ \ \text{and} \ \ h_2 \in H_2(t). \tag{1.24}$$

Proof First of all, for each $t \geq 0$, we let

$$\hat{H}_2(t) \triangleq \mathscr{K}(t)H^C \ \ \text{and} \ \ \hat{H}_1(t) \triangleq (I - \mathscr{K}(t))H^C. \tag{1.25}$$

Let $\hat{\Gamma}$ be the circle $\partial\mathbb{B}(0, (\delta + 1)/2)$ with the anticlockwise direction in $\mathbb{C}^1$. From (1.18), we have

$$\mathscr{K}(t)^2 = \left(\frac{1}{2\pi i}\right)^2 \int\limits_{\hat{\Gamma}} \int\limits_{\Gamma} \frac{-1}{\hat{\lambda} - \lambda} \Big[(\hat{\lambda} I - \mathscr{P}(t)^C)^{-1} - (\lambda I - \mathscr{P}(t)^C)^{-1} \Big] \mathrm{d}\lambda \mathrm{d}\hat{\lambda}.$$

Meanwhile, since $\hat{\Gamma}$ lies outside Γ, we find that

$$\frac{1}{2\pi i} \int\limits_{\Gamma} \frac{1}{\hat{\lambda} - \lambda} \mathrm{d}\lambda = 0 \text{ and } \frac{1}{2\pi i} \int\limits_{\hat{\Gamma}} \frac{1}{\hat{\lambda} - \lambda} \mathrm{d}\hat{\lambda} = 1.$$

From the above three equalities, we find that $\mathscr{K}(t)^2 = \mathscr{K}(t)$. Hence,

$$\mathscr{K}(t) : H^C \to \hat{H}_2(t) \text{ is a projection.} \tag{1.26}$$

Furthermore, one can easily check that

$$H^C = \hat{H}_1(t) \bigoplus \hat{H}_2(t); \tag{1.27}$$

and that both $\hat{H}_1(t)$ and $\hat{H}_2(t)$ are invariant w.r.t. $\mathscr{P}(t)^C$.

Then we prove that $\mathbb{P}(t)$, with $t \geq 0$, is a linear operator from H to H. For this purpose, it suffices to show that

$$\mathscr{K}(t)h \in H, \quad \text{when } h \in H \text{ and } t \geq 0. \tag{1.28}$$

The proof of (1.28) is as follows. By (1.18), we see that

$$\mathscr{K}(t)h = \frac{\delta}{2\pi} \int_0^{2\pi} \left(\delta e^{i\theta} I - \mathscr{P}(t)^C\right)^{-1} e^{i\theta} \mathrm{d}\theta\, h, \quad \text{when } h \in H \text{ and } t \geq 0. \tag{1.29}$$

Write F for the conjugate map from H^C to H^C, i.e., $F(h + ig) = h - ig$ for any $h, g \in H$. We now claim that

$$F\left(\left(\delta e^{i\theta} I - \mathscr{P}(t)^C\right)^{-1} e^{i\theta} h\right) = (\delta e^{-i\theta} I - \mathscr{P}(t)^C)^{-1} e^{-i\theta} h \quad \text{for all } \theta \in [0, 2\pi],\ h \in H,\ t \geq 0. \tag{1.30}$$

When (1.30) is proved, it follows from (1.29) and (1.30) that for all $t \geq 0$ and $h \in H$,

$$F(\mathscr{K}(t)h) = \frac{\delta}{2\pi} \int_0^{2\pi} \left(\delta e^{i\theta} I - \mathscr{P}(t)^C\right)^{-1} e^{i\theta} d\theta\, h = \mathscr{K}(t)h,$$

which leads to (1.28). To show (1.30), we arbitrarily fix $\theta \in [0, 2\pi], t \geq 0$ and $h \in H$. Write

$$(\delta e^{i\theta} I - \mathscr{P}(t)^C)^{-1} e^{i\theta} h = g_1 + i g_2, \quad g_1, g_2 \in H. \tag{1.31}$$

It is clear that $(\delta e^{i\theta} I - \mathscr{P}(t)^C)(g_1 + ig_2) = e^{i\theta} h$. From this, one can directly check the following equality: $(\delta e^{-i\theta} I - \mathscr{P}(t)^C)(g_1 - ig_2) = e^{-i\theta} h$. Hence, we have that

$$(\delta e^{-i\theta} I - \mathscr{P}(t)^C)^{-1}(e^{-i\theta} h) = g_1 - ig_2 = F(g_1 + ig_2).$$

This, along with (1.31), leads to (1.30).

Next we prove that $\mathbb{P}(t)$, with $t \geq 0$, is a projection on H. Let $H_1(t)$ and $H_2(t)$, with $t \geq 0$, be defined by (1.21). One can directly verify that

$$\hat{H}_1(t) = H_1^C(t); \quad \hat{H}_2(t) = H_2^C(t). \tag{1.32}$$

By (1.26) and (1.32), we see that $\mathscr{K}(t)$ (with $t \geq 0$) is a projection from H^C onto $H_2(t)^C$. Thus, for each $t \geq 0$, we have that $\mathbb{P}(t)(h_1 + h_2) = (I - \mathscr{K}(t))(h_1 + h_2) = h_1$, when $h_1 \in H_1(t), h_2 \in H_2(t)$, i.e., $\mathbb{P}(t)$ is a projection from H onto $H_1(t)$. Besides, (1.20) follows from (1.27) and (1.32).

Finally, we will show properties (a)–(f) one by one. *The proof of (a):* Since $\mathscr{P}(\cdot)$ is T-periodic, so is $\mathscr{K}(\cdot)$ (see (1.18)). This, along with (1.19), shows the T-periodicity of $\mathbb{P}(\cdot)$. Then by (1.21), both $H_1(\cdot)$ and $H_2(\cdot)$ are T-periodic. *The proof of (b):* Let $t \geq 0$. Since $\hat{H}_1(t)$ and $\hat{H}_2(t)$ are invariant w.r.t. $\mathscr{P}(t)^C$, so are $H_1(t)^C$ and $H_2(t)^C$ (see (1.32)). Hence, $H_1(t)$ and $H_2(t)$ are invariant w.r.t. $\mathscr{P}(t)$. *The proof of (c):* By (1.17) and (1.15), we see that $\dim H_1(t)^C = n_0$, which leads to (1.22). *The proof of (d) and (e):* Let $0 \leq t \leq s < \infty$. By (1.7), we have that $\Phi(s,t)\mathscr{P}(t) = \mathscr{P}(s)\Phi(s,t)$. By this, one can directly verify that $\Phi(s,t)^C \mathscr{K}(t) = \mathscr{K}(s)\Phi(s,t)^C$, from which, (e) follows at once. Also, this equality, along with (1.21), (1.19) and (1.28), indicates that

$$\Phi(s,t)H_1(t) \subseteq \mathbb{P}(s)H = H_1(s), \tag{1.33}$$

From (1.33), we see that $\Phi(s,t) \in \mathscr{L}(H_1(t), H_1(s))$. Similarly, one can show that $\Phi(s,t) \in \mathscr{L}(H_2(t), H_2(s))$. Hence, (d) stands.

The proof of (f): Since $\sigma\left(\mathscr{P}(t)^C\big|_{H_2(t)^C}\right) \setminus \{0\}$ consists of all such eigenvalues λ that $|\lambda| < 1$, the spectral radius of $\mathscr{P}^C(0)\big|_{H_2(0)^C}$ equals to $\hat{\delta}$. Thus, it follows from the equivalent definition of the spectral radius that $\hat{\delta} = \lim_{k\to\infty} \|\left(\mathscr{P}(0)^C\big|_{H_2(0)^C}\right)^k\|^{\frac{1}{k}}$ (see Theorem 4 on Page 212 in [97]). We arbitrarily fix a $\rho \in (0, \hat{\rho})$. It is clear that $\hat{\delta} \triangleq e^{-\hat{\rho}T} < e^{-\rho T}$. Thus there is positive integer $\hat{N}$ so that $\|\left(\mathscr{P}(0)^C\big|_{H_2(0)^C}\right)^k\| < e^{-\rho kT}$ for all $k \geq \hat{N}$, which implies that

$$\|\left(\mathscr{P}(0)\big|_{H_2(0)}\right)^k\| < e^{-\rho kT} \quad \text{for all} \quad k \geq \hat{N}. \tag{1.34}$$

Notice that $\Phi(\cdot,\cdot)$ is strongly continuous. It follows from the uniform boundedness theorem that $\{\Phi(s,t),\ T \geq s \geq t \geq 0\}$ is bounded. Write

$$C_1 \triangleq \max_{0\leq t_1\leq t_2\leq T} \|\Phi(t_2,t_1)\| \in \mathbb{R}^+ \quad \text{and} \quad C_\rho \triangleq (C_1+1)^2 e^{3\rho T} \in \mathbb{R}^+. \tag{1.35}$$

We are going to show that the above C_ρ satisfies (1.24). For this purpose, we let $0 \le t \le s < \infty$ and $h_2 \in H_2(t)$. For each $r \in \mathbb{R}^+$, we denote by $[r]$ the integer so that $r - 1 < [r] \le r$. There are only two possibilities: (i) $[s/T] = [t/T]$ and (ii) $[s/T] \neq [t/T]$. In the first case, it follows from (1.7) and (1.35) that

$$\begin{aligned}\|\Phi(s,t)h_2\| &= \|\Phi(s-[t/T]T, t-[t/T]T)h_2\| \le \|\Phi(s-[t/T]T, t-[t/T]T)\|\,\|h_2\| \\ &\le C_1\|h_2\| \le C_1 e^{\rho T} e^{-\rho(s-t)}\|h_2\| < (C_1+1)^2 e^{3\rho T} e^{-\rho(s-t)}\|h_2\| = C_\rho e^{-\rho(s-t)}\|h_2\|,\end{aligned}$$

i.e., C_ρ satisfies (1.24) in the first case. In the second case, we have that $[s/T]T \ge [t/T]T + T$; and it follows from (d) and (a) that $\Phi([t/T]T + T, t)h_2 \in H_2(0)$. These, along with (1.7) and (1.34), indicate that as $(s-t)$ is large enough,

$$\begin{aligned}\|\Phi(s,t)h_2\| &\le \left\|\Phi\big(s, [s/T]T\big)\right\| \cdot \left\|\mathscr{P}(0)^{([s/T]-[t/T]-1)}\Phi\big([t/T]T + T,\ t\big)h_2\right\| \\ &\le \left\|\Phi\big(s - [s/T]T,\ 0\big)\right\| \cdot e^{-\rho T([s/T]-[t/T]-1)} \cdot \left\|\Phi\big(T,\ t - [t/T]T\big)\right\|\,\|h_2\|.\end{aligned}$$

By this and (1.35), one can check that C_ρ satisfies (1.24) in this case. This ends the proof. □

The following consequence of Proposition 1.4 provides a useful judgement on the exponential stability of the Eq. (1.1) with the null control.

Corollary 1.1 *Equation (1.1) with the null control is exponentially stable if and only if $\sigma(\mathscr{P}(0)) \in \mathbb{B}$, where $\mathscr{P}(\cdot)$ is the periodic map associated with $A(\cdot)$.*

Proof By the exponential stability of Eq. (1.1) with the null control, there is $C > 0$ and $\delta > 0$ so that $\|y(s; 0, h, 0)\| \le Ce^{-\delta s}\|h\|$ for all $s \ge 0$. From this, it follows that $\|\Phi(nT, 0)\| \le Ce^{-\delta nT}$ for all $n \in \mathbb{N}$. Thus,

$$\lim_{n\to\infty} \sqrt[n]{\|\mathscr{P}(0)^n\|} = \lim_{n\to\infty} \sqrt[n]{\|\Phi(nT, 0)} = e^{-\delta T} < 1.$$

Therefore, the spectral radius of $\mathscr{P}(0)$ is less than 1. Conversely, when the spectral radius of $\mathscr{P}(0)$ is less than 1, it follows by (1.17), (1.19) and (1.21) that $n_0 = 0$, $\mathbb{P}(0) = 0$ and $H_1(0) = \{0\}$. These, along with (f) of Proposition 1.4, shows that Eq. (1.1) with the null control is exponentially stable. This ends the proof. □

We now introduce some properties on $\mathscr{P}(0)^{*C}$, which equals to the adjoint operator of $\mathscr{P}(0)^C$. It is clear that

$$\sigma\big(\mathscr{P}(0)^{*C}\big) = \overline{\sigma\big(\mathscr{P}(0)^C\big)}. \tag{1.36}$$

Write $\bar{l}_j$ for the algebraic multiplicity of $\bar{\lambda}_j$ w.r.t. $\mathscr{P}(0)^{*C}$. Then

$$\bar{l}_j = l_j \text{ for all } j, \text{ and } \bar{l}_1 + \cdots + \bar{l}_{\hat{n}} = n_0, \tag{1.37}$$

where $\hat{n}$ and n_0 are given by (1.15) and (1.17) respectively. Let Γ be the circle used to define $\mathscr{K}(\cdot)$ (see (1.18)). The Kato projection with respect to $\mathscr{P}^{*C}$ reads:

$$\tilde{\mathscr{K}} = \frac{1}{2\pi i}\int_{\Gamma}(\lambda I - \mathscr{P}(0)^{*C})^{-1}d\lambda. \tag{1.38}$$

Thus we find that

$$H^C = \widehat{\widetilde{H}}_1 \bigoplus \widehat{\widetilde{H}}_2, \quad \text{with } \widehat{\widetilde{H}}_1 \triangleq (I - \tilde{\mathscr{K}})H^C,\ \widehat{\widetilde{H}}_2 \triangleq \tilde{\mathscr{K}}H^C, \tag{1.39}$$

and

$$\text{both } \widehat{\widetilde{H}}_1 \text{ and } \widehat{\widetilde{H}}_2 \text{ are invariant w.r.t. } \mathscr{P}^{*C}. \tag{1.40}$$

Let

$$\tilde{\mathbb{P}} \triangleq (I - \tilde{\mathscr{K}})\big|_H . \tag{1.41}$$

Proposition 1.5 *Let $\tilde{\mathbb{P}}$ be defined by (1.41). Then, $\tilde{\mathbb{P}}$ is a projection on H; $H = \tilde{H}_1 \bigoplus \tilde{H}_2$, where $\tilde{H}_1 \triangleq \tilde{\mathbb{P}}H$ and $\tilde{H}_2 \triangleq (I - \tilde{\mathbb{P}})H$; $\mathscr{P}^*\tilde{H}_1 \subseteq \tilde{H}_1$; $\sigma\big(\mathscr{P}^{*C}|_{\tilde{H}_1^C}\big) = \{\bar{\lambda}_j\}_{j=1}^{\hat{n}}$ and $\sigma\big(\mathscr{P}^{*C}|_{\tilde{H}_2^C}\big) \subseteq \mathbb{B}$; and $\dim\tilde{H}_1 = n_0$. It further holds that*

$$\tilde{\mathbb{P}} = \mathbb{P}(0)^*; \tag{1.42}$$

$$\tilde{H}_1 = \mathbb{P}(0)^*H = \mathbb{P}(0)^*H_1, \quad \textit{where } H_1 \textit{ is given by } (1.21); \tag{1.43}$$

$$\xi \in \tilde{H}_1^C, \quad \textit{when } \mu \in \sigma\big(\mathscr{P}(0)^{*C}\big) \setminus \mathbb{B} \textit{ and } (\mu I - \mathscr{P}(0)^{*C})\xi = 0. \tag{1.44}$$

Proof By (1.36)–(1.40), one can make use of the exactly same way utilized in the proof of Proposition 1.4 to verify all properties in Proposition 1.5, except for (1.42)–(1.44). Since $(\bar{\lambda}I - \mathscr{P}(0)^{*C})^{-1} = \big((\lambda I - \mathscr{P}(0)^C)^{-1}\big)^*$, (1.42) follows from (1.38), (1.18) and (1.19). Now, we prove (1.43). The first equation of (1.43) follows from the definition of $\tilde{H}_1$ and (1.42). It is clear that $\mathbb{P}(0)^*H \supseteq \mathbb{P}(0)^*H_1$. On the other hand, since

$$\mathbb{P}(0)^*\mathbb{P}(0)h = 0 \Rightarrow \langle h, \mathbb{P}(0)^*\mathbb{P}(0)h\rangle = 0 \Rightarrow \mathbb{P}(0)h = 0,$$

we see that $\mathscr{N}(\mathbb{P}(0)^*\mathbb{P}(0)) \subseteq \mathscr{N}(\mathbb{P}(0))$. Since $H_1(0) = \mathbb{P}(0)H$ (see (1.20) and (1.21)), we find

$$\begin{aligned}\mathbb{P}(0)^*H_1(0) &= \mathbb{P}(0)^*\mathbb{P}(0)H = \mathscr{R}(\mathbb{P}(0)^*\mathbb{P}(0))\\ &= \mathscr{N}(\mathbb{P}(0)^*\mathbb{P}(0))^\perp \supseteq \mathscr{N}(\mathbb{P}(0))^\perp = \mathscr{R}(\mathbb{P}(0)^*) = \mathbb{P}(0)^*H,\end{aligned}$$

which leads to (1.43).

The proof of (1.44) is as follows. Since $\mathscr{P}(0)^{*C}\xi = \mu\xi$, we derive from (1.38) that

$$\tilde{\mathscr{K}}\xi = \frac{1}{2\pi i}\int_{\Gamma} \left(\lambda I - \mathscr{P}(0)^{*C}\right)^{-1} \mathrm{d}\lambda\, \xi = \frac{1}{2\pi i}\int_{\Gamma} (\lambda - \mu)^{-1}\,\mathrm{d}\lambda\, \xi = 0.$$

Hence, $\xi \in (I - \tilde{\mathscr{K}})H^C$. Meanwhile, by the definitions of $\tilde{\mathbb{P}}$ and $\tilde{H}_1$, we find that

$$\tilde{H}_1^C = \left((I - \tilde{\mathscr{K}})\big|_H\right)^C H^C = (I - \tilde{\mathscr{K}})H^C.$$

Thus, we see that $\xi \in \tilde{H}_1^C$. This completes the proof. □

We end this section with introducing some preliminaries on the linear algebra.

Definition 1.6 Given a matrix $M \in \mathbb{R}^{m\times k}$, with $m, k \in \mathbb{N}$, its Moore-Penrose inverse is the matrix $M^\dagger \in \mathbb{R}^{k\times m}$ which satisfies the following conditions:

$$\begin{array}{ll} (i)\ \ MM^\dagger M = M; & (ii)\ \ M^\dagger M M^\dagger = M^\dagger; \\ (iii)\ \ (MM^\dagger)^* = MM^\dagger; & (iv)\ \ (M^\dagger M)^* = M^\dagger M. \end{array} \tag{1.45}$$

Remark 1.1 It deserves to mention that each matrix $M \in \mathbb{R}^{m\times k}$ has a unique Moore-Penrose inverse $M^\dagger$ (see [73]).

Lemma 1.1 *Let M be a symmetric matrix in $\mathbb{R}^{n\times n}$. Then, any vector ξ in $\mathbb{R}^n$ can be decomposed into two orthogonal vectors $M^\dagger M\xi$ and $\xi - M^\dagger M\xi$ so that*

$$M^\dagger M\xi \in \mathscr{R}(M) \ \ \textit{and} \ \ \xi - M^\dagger M\xi \in \mathscr{N}(M). \tag{1.46}$$

Proof From the definition of the Moore-Penrose inverse, one can easily check that $(\hat{M}^*)^\dagger = (\hat{M}^\dagger)^*$ for any $\hat{M} \in \mathbb{R}^{n\times n}$. This, together with the symmetry of M, implies that $M^\dagger$ is also symmetric. Thus, from (iii) in (1.45), we find that $M^\dagger M = MM^\dagger$. Hence,

$$M^\dagger M\xi = (M^\dagger M)\xi = (MM^\dagger)\xi = M(M^\dagger\xi) \in \mathscr{R}(M);$$

and

$$M(\xi - M^\dagger M\xi) = M\xi - (MM^\dagger M)\xi = M\xi - M\xi = 0.$$

These lead to (1.46). Finally, by the symmetry of M, we see that $\mathscr{R}(M) = \mathscr{N}(M^*)^\perp = \mathscr{N}(M)^\perp$. Hence, the vectors $M^\dagger M\xi$ and $(\xi - M^\dagger M\xi)$ are orthogonal. This ends the proof. □

The following lemma is a fundamental result of linear algebra (see Theorem 1 on Page 67 in [40]). We omit its proof.

Lemma 1.2 *Let $\hat{H}$ be a real linear space with* $\dim \hat{H} < \infty$. *Let L be a linear map on $\hat{H}$. Then $\hat{H}$ can be uniquely decomposed as*

$$\hat{H} = \hat{H}_1(L) \bigoplus \hat{H}_2(L), \tag{1.47}$$

where $\hat{H}_1(L)$ and $\hat{H}_2(L)$ are invariant under L and satisfy respectively

$$\sigma\left(L\big|_{\hat{H}_1(L)}\right) \subset \mathbb{B} \quad and \quad \sigma\left(L\big|_{\hat{H}_2(L)}\right) \subset \mathbb{B}^c. \tag{1.48}$$

Remark 1.2 Let L be a linear map on $\mathbb{R}^n$. Let $V \subset \mathbb{R}^n$ be an invariant subspace of L. Then, Lemma 1.2 provides a unique decomposition of V corresponding to the map $L\big|_V$. We will simply denote this decomposition by $(V_1(L), V_2(L))$, when there is no risk to cause any confusion.

Lemma 1.3 *Let Z be a finite dimensional vector space and L be a linear map on Z. Suppose that $Y \subseteq Z$ is an invariant subspace of L. Then*

$$Y_1(L) \subseteq Z_1(L) \quad and \quad Y_2(L) \subseteq Z_2(L). \tag{1.49}$$

Proof Let $\hat{Z} = Y_1(L) \bigoplus Z_2(L)$. Then

$$Z = \hat{Z} + Z = (Y_1(L) + Z_1(L)) + Z_2(L). \tag{1.50}$$

Subspaces $Y_1(L)$ and $Z_1(L)$ are invariant under L, so is $Y_1(L) + Z_1(L)$. It is clear that $\sigma\left(L\big|_{Y_1(L)+Z_1(L)}\right) \subset \mathbb{B}$ and $\sigma\left(L\big|_{Z_2(L)}\right) \subset \mathbb{B}^c$. These lead to $(Y_1(L) + Z_1(L)) \bigcap Z_2(L) = \{0\}$. Along with (1.50), this yields that $Z = (Y_1(L) + Z_1(L)) \bigoplus Z_2(L)$. By the uniqueness of in Lemma 1.2, we see that $Y_1(L) + Z_1(L) = Z_1(L)$, which implies that $Y_1(L) \subseteq Z_1(L)$. Similarly, we can verify the second conclusion in (1.49). This ends the proof. □

1.2 Linear Quadratic Optimal Control Problems

Optimal control problems, where controlled systems are linear equations and the cost functionals have quadratic forms, are referred to as LQ problems. The LQ theory is one of the most important subjects in the control field. It plays a very important role in the studies of the stabilization. In this section, we introduce some properties on some LQ problems, associated with the pair $[A(\cdot), B(\cdot)]$, via two cases which are the finite horizon case and the infinite horizon case. For more materials about this subject, we refer readers to [34].

1.2.1 Finite Horizon Case

In this subsection, we study some LQ problems in finite horizons. First of all, we introduce some notations. Given a real Hilbert space $\hat{H}$, we write

$$S\mathscr{L}(\hat{H}) \triangleq \left\{O \in \mathscr{L}(\hat{H}) \mid O^* = O\right\};\ S\mathscr{L}^+(\hat{H}) \triangleq \left\{O \in S\mathscr{L}(\hat{H}) \mid \langle h, Oh\rangle \geq 0,\ \forall h \in \hat{H}\right\}.$$

By $O \gg 0$, we mean that $O - \delta I \in S\mathscr{L}^+(\hat{H})$ for some $\delta > 0$. To introduce our LQ problems, we let $\hat{T} > 0$ be arbitrarily fixed. For each $t \in [0, \hat{T})$, let $\mathscr{U}_t^{\hat{T}} \triangleq L^2(t, \hat{T}; U)$. For each $t \in [0, \hat{T})$, $h \in H$ and $u \in \mathscr{U}_t^{\hat{T}}$, we write $y(\cdot; t, h, u, \hat{T})$ for the mild solution to the equation:

$$y'(s) = A(s)y(s) + B(s)u(s),\quad s \in [t, \hat{T}];\quad y(t) = h. \tag{1.51}$$

Then for each $t \in [0, \hat{T})$ and each $h \in H$, we define the following cost functional:

$$J_{t,h}^{\hat{T}}(u) = \int_t^{\hat{T}} \Big[\langle y(s), Qy(s)\rangle + \langle u(s), Ru(s)\rangle\Big]\mathrm{d}s + \langle y(\hat{T}), My(\hat{T})\rangle,\quad u \in \mathscr{U}_t^{\hat{T}}, \tag{1.52}$$

with $y(\cdot) \triangleq y(\cdot; t, h, u, \hat{T})$. Here and throughout this section, we assume that

$$Q,\ M \in S\mathscr{L}^+(H),\quad R \in S\mathscr{L}^+(U),\quad R \gg 0. \tag{1.53}$$

Then we define the following LQ problem:

$$(LQ)_{t,h}^{\hat{T}}:\qquad W^{\hat{T}}(t, h) \triangleq \inf_{u \in \mathscr{U}_t^{\hat{T}}} J_{t,h}^{\hat{T}}(u),\quad t \in [0, \hat{T}),\ h \in H. \tag{1.54}$$

Definition 1.7 (i) The map $(t, h) \to W^{\hat{T}}(t, h)$, $(t, h) \in [0, \hat{T}) \times H$, is called the value function associated with the family $\{(LQ)_{t,h}^{\hat{T}}\}_{(t,h)\in[0,\hat{T})\times H}$. It can be extended to a function over $[0, \hat{T}] \times H$ by setting $W^{\hat{T}}(\hat{T}, h) \triangleq \langle h, Mh\rangle$. (ii) Given $(t, h) \in [0, \hat{T}) \times H$, the problem $(LQ)_{t,h}^{\hat{T}}$ is solvable if there exists a control $\bar{u} \in \mathscr{U}_t^{\hat{T}}$ so that $J_{t,h}^{\hat{T}}(\bar{u}) = W^{\hat{T}}(t, h)$. Such a control $\bar{u}$ is called an optimal control, the solution $\bar{y}(\cdot) \triangleq y(\cdot; t, h, \bar{u}, \hat{T})$ is called the corresponding optimal trajectory, while the pair $(\bar{y}, \bar{u})$ is referred to as an optimal pair to the problem $(LQ)_{t,h}^{\hat{T}}$.

One can easily verify that each $(LQ)_{t,h}^{\hat{T}}$ has a unique optimal control. So it is solvable. For each $t \in [0, \hat{T})$, we introduce the following operators: Define $G_t^{\hat{T}} : \mathscr{U}_t^{\hat{T}} \to \mathscr{H}_t^{\hat{T}} \triangleq L^2(t, \hat{T}; H)$ by

$$(G_t^{\hat{T}}(u))(s) \triangleq \int_t^s \Phi(s, r)B(r)u(r)\mathrm{d}r,\quad s \in [t, \hat{T}],\quad \text{for each}\ \ u \in \mathscr{U}_t^{\hat{T}};$$

Define $\hat{G}_t^{\hat{T}} : \mathscr{U}_t^{\hat{T}} \to H$ by setting

$$\hat{G}_t^{\hat{T}}(u) \triangleq \int_t^{\hat{T}} \Phi(\hat{T}, r)B(r)u(r)\mathrm{d}r\ \ \text{for each}\ \ u \in \mathscr{U}_t^{\hat{T}};$$

Define $\Phi_t^{\hat{T}} : H \to \mathscr{H}_t^{\hat{T}}$ by setting

$$(\Phi_t^{\hat{T}}(h))(s) \triangleq \Phi(s,t)h, \ s \in [t, \hat{T}], \quad \text{for each } \ h \in H;$$

Define $\hat{\Phi}_t^{\hat{T}} : H \to H$ by setting

$$\hat{\Phi}_t^{\hat{T}}(h) \triangleq \Phi(\hat{T}, t)h \ \text{ for each } \ h \in H;$$

Define $Q_t^{\hat{T}} : \mathscr{H}_t^{\hat{T}} \to \mathscr{H}_t^{\hat{T}}$ by setting

$$(Q_t^{\hat{T}}(y))(s) \triangleq Qy(s), \ s \in [t, \hat{T}], \quad \text{for each } \ y \in \mathscr{H}_t^{\hat{T}};$$

Define $R_t^{\hat{T}} : \mathscr{U}_t^{\hat{T}} \to \mathscr{U}_t^{\hat{T}}$ by setting

$$(R_t^{\hat{T}}(u))(s) \triangleq Ru(s), \quad \text{a.e. } s \in [t, \hat{T}], \quad \text{for each } \ u \in \mathscr{U}_t^{\hat{T}};$$

One can easily check that all operators above are linear and bounded. Furthermore, one can verify that for each $t \in [0, \hat{T})$, $h \in H$ and $u \in \mathscr{U}_t^{\hat{T}}$,

$$J_{t,h}^{\hat{T}}(u(\cdot)) = \langle u, \Gamma_{t,1}(u)\rangle_{\mathscr{U}_t^{\hat{T}}} + 2\langle u, \Gamma_{t,2}(h)\rangle_{\mathscr{U}_t^{\hat{T}}} + \langle h, \Gamma_{t,3}(h)\rangle_{\mathscr{H}_t^{\hat{T}}}, \tag{1.55}$$

where

$$\Gamma_{t,1} \triangleq (G_t^{\hat{T}})^* Q_t^{\hat{T}} G_t^{\hat{T}} + R_t^{\hat{T}} + (\hat{G}_t^{\hat{T}})^* M \hat{G}_t^{\hat{T}} \in \mathscr{L}(\mathscr{U}_t^{\hat{T}}), \tag{1.56}$$

$$\Gamma_{t,2} \triangleq (G_t^{\hat{T}})^* Q_t^{\hat{T}} \Phi_t^{\hat{T}} + (\hat{G}_t^{\hat{T}})^* M \hat{\Phi}_t^{\hat{T}} \in \mathscr{L}(H, \mathscr{U}_t^{\hat{T}}), \tag{1.57}$$

$$\Gamma_{t,3} \triangleq (\Phi_t^{\hat{T}})^* Q_t^{\hat{T}} \Phi_t^{\hat{T}} + (\hat{\Phi}_t^{\hat{T}})^* M \hat{\Phi}_t^{\hat{T}} \in \mathscr{L}(H). \tag{1.58}$$

Because $R \gg 0$ in $S\mathscr{L}^+(U)$, we find that $R_t^{\hat{T}} \gg 0$. Then by (1.56), we see that $\Gamma_{t,1} \gg 0$. From (1.55) to (1.58), one can easily obtain the following proposition:

Proposition 1.6 *Suppose that (1.53) holds. Let $\Gamma_{t,1}$, $\Gamma_{t,2}$ and $\Gamma_{t,3}$ are given by (1.56)–(1.58), respectively. Then the following assertions hold:*

(i) For each $t \in [0, \hat{T})$ and $h \in H$, $\bar{u}_{t,h}$ is the optimal control to Problem $(LQ)_{t,h}^{\hat{T}}$ if and only if

$$\bar{u}_{t,h} = -\Gamma_{t,1}^{-1}\Gamma_{t,2}h. \tag{1.59}$$

(ii) For each $t \in [0, \hat{T})$ and $h \in H$,

$$W^{\hat{T}}(t,h) = \langle h, (\Gamma_{t,3} - \Gamma_{t,2}^*\Gamma_{t,1}^{-1}\Gamma_{t,2})h\rangle. \tag{1.60}$$

To construct a feedback optimal control of the problem $(LQ)_{t,h}^{\hat{T}}$, we need the help of the following Riccati equation (whose solution is an operator-valued function):

$$\begin{cases} \dot{\Upsilon}(t) + A(t)^*\Upsilon(t) + \Upsilon(t)A(t) + Q - \Upsilon(t)B(t)R^{-1}B(t)^*\Upsilon(t) = 0, \ t \in (0, \hat{T}), \\ \Upsilon(\hat{T}) = M. \end{cases} \tag{1.61}$$

To study the above equation, we first introduce some concepts stated in the following definition:

Definition 1.8 Let $-\infty < a < b < \infty$. (i) $C([a,b]; S\mathscr{L}(H))$ denotes the space of all continuous functions from $[a,b]$ to $S\mathscr{L}(H)$, endowed with the norm:

$$\|f\|_{C([a,b];S\mathscr{L}(H))} \triangleq \sup_{t\in[a,b]} \|f(t)\|_{\mathscr{L}(H)}, \quad f \in C([a,b]; S\mathscr{L}(H)).$$

(ii) $C_s([a,b]; S\mathscr{L}(H))$ denotes the vector space of all strongly continuous functions from $[a,b]$ to $S\mathscr{L}(H)$. (iii) $C_u([a,b]; S\mathscr{L}(H))$ is the space $C_s([a,b]; S\mathscr{L}(H))$ endowed with the norm:

$$\|f\|_{C_u([a,b];S\mathscr{L}(H))} \triangleq \sup_{t\in[a,b]} \|f(t)\|_{\mathscr{L}(H)}. \tag{1.62}$$

(iv) The sequence $\{f_n\}$ is said to converge strongly to f in $C_s([a,b]; S\mathscr{L}(H))$, denoted by $\lim_{n\to\infty} f_n = f$, in $C_s([a,b]; S\mathscr{L}(H))$, if for each $h \in H$, $f_n(s)h \to f(s)h$ in H uniformly w.r.t. $s \in [a,b]$.

Notice that the quantity on the right hand side of (1.62) is finite because of the uniform boundedness principle, and that both $C([a,b]; S\mathscr{L}(H))$ and $C_u([a,b]; S\mathscr{L}(H))$ are Banach spaces.

Definition 1.9 We present two types of solutions to Eq. (1.61).

(i) A mild solution of Eq. (1.61) is a function $\Upsilon \in C_u([0,\hat{T}]; S\mathscr{L}(H))$ so that for each $s \in [0,\hat{T}]$ and $h \in H$,

$$\begin{aligned} \Upsilon(s)h = \Phi(\hat{T},s)^*M\Phi(\hat{T},s)h + \int_s^{\hat{T}} \Phi(\sigma,s)^*Q\Phi(\sigma,s)h\mathrm{d}\sigma \\ - \int_s^{\hat{T}} \Phi(\sigma,s)^*\Upsilon(\sigma)B(\sigma)R^{-1}B(\sigma)^*\Upsilon(\sigma)\Phi(\sigma,s)h\mathrm{d}\sigma. \end{aligned} \tag{1.63}$$

(ii) A weak solution of Eq. (1.61) is a function $\Upsilon \in C_u([0,\hat{T}]; S\mathscr{L}(H))$ so that $\Upsilon(\hat{T}) = M$ and so that for any $h_1, h_2 \in \mathscr{D}(A)$, the function $s \to \langle h_1, \Upsilon(s)h_2\rangle$ is differentiable in $[0,\hat{T}]$ and verifies the equation:

$$\begin{aligned} \frac{d}{ds}\langle h_1, \Upsilon(s)h_2\rangle + \langle A(s)h_1, \Upsilon(s)h_2\rangle + \langle \Upsilon(s)h_1, A(s)h_2\rangle + \langle h_1, Qh_2\rangle \\ -\langle B(s)^*\Upsilon(s)h_1, R^{-1}B(s)^*\Upsilon(s)h_2\rangle = 0, \quad s \in [0,\hat{T}]. \end{aligned} \tag{1.64}$$

Proposition 1.7 *Let $\Upsilon \in C_u([0, \hat{T}]; S\mathscr{L}(H))$. Then Υ is a mild solution of Eq. (1.61) if and only if Υ is a weak solution of Eq. (1.61).*

Proof Let Υ be a mild solution of (1.61). Then for any $h_1, h_2 \in H$ and $s \in [0, \hat{T}]$, we have that

$$\begin{aligned}\langle h_1, \Upsilon(s)h_2\rangle &= \langle \Phi(\hat{T}, s)h_1,\ M\Phi(\hat{T}, s)h_2\rangle + \int_s^{\hat{T}} \langle \Phi(\sigma, s)h_1,\ Q\Phi(\sigma, s)h_2\rangle d\sigma \\ &\quad - \int_s^{\hat{T}} \langle B(\sigma)^*\Upsilon(\sigma)\Phi(\sigma, s)h_1,\ R^{-1}B(\sigma)^*\Upsilon(\sigma)\Phi(\sigma, s)h_2\rangle d\sigma.\end{aligned}$$

From this, we find that when $h_1, h_2 \in \mathscr{D}(A)$, the function $s \to \langle h_1, \Upsilon(s)h_2\rangle$ is differentiable with respect to s over $[0, \hat{T}]$ and satisfies (1.64). Conversely, if Υ is a weak solution of (1.61), then one can verify that when $h_1, h_2 \in \mathscr{D}(A)$ and $s \in [0, \hat{T}]$,

$$\begin{aligned}&\frac{d}{ds}\langle \Phi(\hat{T}, s)h_1,\ \Upsilon(s)\Phi(\hat{T}, s)h_2\rangle \\ &= \langle \Phi(\hat{T}, s)h_1,\ Q\Phi(\hat{T}, s)h_2\rangle - \langle \Phi(\hat{T}, s)h_1,\ \Upsilon(s)B(s)R^{-1}B(s)^*\Upsilon(s)\Phi(\hat{T}, s)h_2\rangle.\end{aligned}$$

Integrating the above from 0 to $\hat{T}$ leads to (1.63) with $h \in \mathscr{D}(A)$. Then the density of $\mathscr{D}(A)$ in H yields the desired result. This completes the proof. □

Theorem 1.1 *Suppose that (1.53) holds. Then the Eq. (1.61) has a unique mild solution $\Upsilon \in C_u([0, \hat{T}]; S\mathscr{L}(H))$, with $\Upsilon(s) \geq 0$ for each $s \in [0, \hat{T}]$.*

Proof We organize the proof by two steps.
Step 1. To show the existence

We first use the contractive mapping theorem to show the local existence. Let

$$C \triangleq \sup_{s\in[0,\hat{T}]} \|e^{sA}\| + \sup_{0\leq t\leq s\leq \hat{T}} \|\Phi(s, t)\| > 1. \tag{1.65}$$

Set

$$\rho \triangleq 2C^2\beta, \quad \text{with } \beta \triangleq C^2(\|M\| + \hat{T}\|Q\|). \tag{1.66}$$

Choose $\tau \in (0, \hat{T})$ so that $(\hat{T} - \tau)(\|Q\| + \rho^2\|BR^{-1}B\|) \leq \beta$ and $2\rho(\hat{T} - \tau)C^2\| BR^{-1}B\| \leq 1/2$. Let

$$\mathscr{B}_{\rho,\tau} \triangleq \left\{ f \in C_u([\hat{T} - \tau, \hat{T}]; S\mathscr{L}(H)) \;\middle|\; \|f\|_{C_u([\hat{T}-\tau,\hat{T}];S\mathscr{L}(H))} \leq \rho \right\}. \tag{1.67}$$

We define a map $F : \mathscr{B}_{\rho,\tau} \to \mathscr{B}_{\rho,\tau}$ in the following manner: for each $h \in H$,

$$F(\Upsilon)(s)h \triangleq \Phi(\hat{T}, s)^* M \Phi(\hat{T}, s)h + \int_s^{\hat{T}} \Phi(\sigma, s)^* Q \Phi(\sigma, s)h d\sigma$$
$$- \int_s^{\hat{T}} \Phi(\sigma, s)^* \Upsilon(\sigma) B(\sigma) R^{-1} B(\sigma)^* \Upsilon(\sigma) \Phi(\sigma, s)h d\sigma, \quad s \in [\hat{T} - \tau, \hat{T}]. \tag{1.68}$$

One can directly check that F is a contractive mapping on $\mathscr{B}_{\rho,\tau}$. Hence, Eq. (1.61) has a unique mild solution $P \in \mathscr{B}_{\rho,\tau}$.

We next prove that $\Upsilon(s) \geq 0$ for any $s \in [\hat{T} - \tau, \hat{T}]$. To this end, notice that Υ is the solution of the following linear equation in $[\hat{T} - \tau, \hat{T}]$:

$$\Upsilon' + L^* \Upsilon + \Upsilon L + Q = 0, \qquad \Upsilon(\hat{T}) = M,$$

where $L(\cdot) \triangleq A(\cdot) - \frac{1}{2} B R^{-1} B^* P$. Denote by $U(s, t)$, $\hat{T} - \tau \leq t \leq s \leq \hat{T}$, the evolution generated by $L(\cdot)$. Then, $\partial_s U(s, t) = L(s) U(s, t)$ and $U(t, t) = I$, when $\hat{T} - \tau \leq t \leq s \leq \hat{T}$. Thus, we have

$$\Upsilon(s) = U(\hat{T}, s)^* M U(\hat{T}, s) + \int_s^{\hat{T}} U(\tau, s)^* Q U(\tau, s) d\tau, \; s \in [0, \hat{T}],$$

from which, it follows that $\Upsilon(s) \geq 0$.

We now claim that $\|\Upsilon(s)\| \leq \beta$ for any $s \in [\hat{T} - \tau, \hat{T}]$. In fact, we have that for each $s \in [\hat{T} - \tau, \hat{T}]$ and $h \in H$,

$$\langle h, \Upsilon(s)h \rangle = \langle \Phi(\hat{T}, s)h, \; M \Phi(\hat{T}, s)h \rangle + \int_s^{\hat{T}} \langle \Phi(r, s)h, \; Q \Phi(r, s)h \rangle dr$$
$$- \int_s^{\hat{T}} \langle \Upsilon(r) \Phi(r, s)h, \; B(r) R^{-1} B(r)^* \Upsilon(r) \Phi(r, s)h \rangle dr \leq \beta \|h\|^2.$$

Since $\Upsilon(s) \geq 0$, the above leads to the desired claim. With the aid of this claim, we can apply the contractive mapping theorem to extend Υ over $[\hat{T} - 2\tau, \hat{T} - \tau]$ so that $\|\Upsilon(s)\| \leq \rho$ for each $s \in [\hat{T} - 2\tau, \hat{T} - \tau]$. Then we can repeat the previous argument step by step to get the desired existence.

Step 2. To show the uniqueness

Let Υ and $\hat{\Upsilon}$ be two mild solutions of the Eq. (1.61). Recall (1.62). Set

$$\hat{\beta} \triangleq \max \left\{ \|\Upsilon(\cdot)\|_{C_u([0,\hat{T}]; S\mathscr{L}(H))}, \; \|\hat{\Upsilon}(\cdot)\|_{C_u([0,\hat{T}]; S\mathscr{L}(H))} \right\}.$$

Choose $\hat{\rho} > 0$ and $\hat{\tau} \in [0, \hat{T}]$ so that

$$\hat{\rho} = 2C^2 \hat{\beta}, \;\; (\hat{T} - \hat{\tau})(\|Q\| + \hat{\rho}^2 \|B R^{-1} B\|) \leq \hat{\beta}, \;\; 2\rho(\hat{T} - \hat{\tau}) C^2 \|B R^{-1} B\| \leq 1/2.$$

It is clear that $\|\Upsilon(s)\| \leq \hat{\rho}$ and $\|\hat{\Upsilon}(s)\| \leq \hat{\rho}$ for all $s \in [0, \hat{T}]$. Define $\mathscr{B}_{\hat{\rho},\hat{\tau}}$ and F by (1.67) and (1.68) with $\rho = \hat{\rho}$ and $\tau = \hat{\tau}$. One can directly check that this new

F is still a contractive mapping on the new ball $\mathscr{B}_{\hat{\rho},\hat{\tau}}$. Hence, $\Upsilon(t) = \hat{\Upsilon}(t)$ for any $t \in [\hat{T} - \hat{\tau}, \hat{T}]$. Moreover, the choice of $\hat{\beta}$ allows us to repeat the above argument in the interval $[\hat{T} - 2\hat{\tau}, \hat{T} - \hat{\tau}]$ and so on.

In summary, we end the proof. □

The identity in the following proposition presents a connection of the functional $J_{t,h}^{\hat{T}}$ with the solution $y(\cdot; t, h, u, \hat{T})$ of the Eq. (1.51). This identity is quoted from [29] where the author proved it through using the Yosida approximation to Riccati equation, while we provide a direct proof here.

Proposition 1.8 *Suppose that (1.53) holds. Then for each $t \in [0, \hat{T}]$, $h \in H$, and $u \in \mathscr{U}_t^{\hat{T}}$,*

$$J_{t,h}^{\hat{T}}(u) = \int_t^{\hat{T}} \left\| R^{1/2}\left(u(s) + R^{-1}B(s)^*\Upsilon(s)y(s; t, h, u, \hat{T})\right)\right\|^2 \mathrm{d}s + \langle h, \ \Upsilon(t)h\rangle, \tag{1.69}$$

where Υ is the mild solution of the Riccati equation (1.61).

Proof Let $h \in H$, $t \in [0, \hat{T}]$ and $u \in \mathscr{U}_t^{\hat{T}}$. Write $y(\cdot) \triangleq y(\cdot; t, h, u, \hat{T})$. From Proposition 1.2,

$$y(s) = \Phi(s,t)h + \int_t^s \Phi(s,r)B(r)u(r)\mathrm{d}r, \quad s \in [t, \hat{T}]. \tag{1.70}$$

By (1.52) and (1.63), after some direct computations, we find that

$$\begin{aligned}\int_t^{\hat{T}} &\left\| R^{1/2}\left(u(s) + R^{-1}B(s)^*\Upsilon(s)y(s; t, h, u)\right)\right\|^2 \mathrm{d}s \\ &= \int_t^T \langle u(s), Ru(s)\rangle \mathrm{d}s + I_1 + I_2 + I_3 + I_4,\end{aligned} \tag{1.71}$$

where

$$I_1 \triangleq \int_t^{\hat{T}} \langle B^*(s)\Upsilon(s)\Phi(s,t)h, \ R^{-1}B^*(s)\Upsilon(s)\Phi(s,t)h\rangle \mathrm{d}s,$$

$$I_2 \triangleq \int_t^{\hat{T}} \left\langle B^*(s)\Upsilon(s)\int_t^s \Phi(s,r)B(r)u(r)\mathrm{d}r, \ R^{-1}B^*(s)\Upsilon(s)\int_t^s \Phi(s,r)B(r)u(r)\mathrm{d}r\right\rangle \mathrm{d}s,$$

$$I_3 \triangleq 2\int_t^{\hat{T}} \left\langle u(s), B^*(s)\Upsilon(s)\left[\Phi(s,t)h + \int_t^s \Phi(s,r)B(r)u(r)\mathrm{d}r\right]\right\rangle \mathrm{d}s,$$

$$I_4 \triangleq 2\int_t^{\hat{T}} \left\langle B^*(s)\Upsilon(s)\Phi(s,t)h, \ R^{-1}B^*(s)\Upsilon(s)\int_t^s \Phi(s,r)B(r)u(r)\mathrm{d}r\right\rangle \mathrm{d}s.$$

From (1.70), we see that

$$I_3 = 2\int_t^{\hat{T}} \langle y(s),\ \Upsilon(s)B(s)u(s)\rangle\, \mathrm{d}s. \tag{1.72}$$

Write

$$\Psi_1(t) \triangleq \Phi(\hat{T},t)^* M\Phi(\hat{T},t) + \int_t^{\hat{T}} \Phi(s,t)^* Q\Phi(s,t)\mathrm{d}s - \Upsilon(t),\ t \in [0,\hat{T}], \tag{1.73}$$

$$\Psi_2(t) \triangleq \int_t^{\hat{T}} \Phi(s,t)^* \Upsilon(s)B(s)R^{-1}B^*(s)\Upsilon(s)\Phi(s,t)\mathrm{d}s,\ t \in [0,\hat{T}]. \tag{1.74}$$

Here, the integral on the right hand side of (1.73) is treated as an operator on H via

$$\Big(\int_t^{\hat{T}} \Phi(s,t)^* Q\Phi(s,t)\mathrm{d}s\Big)h \triangleq \int_t^{\hat{T}} \Phi(s,t)^* Q\Phi(s,t)h\mathrm{d}s,\quad h \in H.$$

The same is said about the term on the right hand side of (1.74). From (1.63), (1.73) and (1.74), we find that

$$\Psi_2(t)h = \Psi_1(t)h \ \text{ for all } \ t \geq 0,\quad h \in H. \tag{1.75}$$

By (1.75) and (1.73), we have that

$$I_1 = \langle h, \Psi_1(t)h\rangle\,. \tag{1.76}$$

To deal with the term I_2, we first observe that $F(r,\sigma) = F(\sigma,r)$ for all $r,\sigma \in [0,s]$ with any fixed $s > 0$, where

$$F(r,\sigma) \triangleq \big\langle \Phi(s,r)B(r)u(r),\ \Upsilon(s)B(s)R^{-1}B^*(s)\Upsilon(s)\Phi(s,\sigma)B(\sigma)u(\sigma)\big\rangle.$$

Thus, we see that

$$I_2 = 2\int_t^{\hat{T}} \mathrm{d}s \int_t^s \mathrm{d}\sigma \int_t^{\sigma} F(r,\sigma)\mathrm{d}r. \tag{1.77}$$

From (1.77), Fubini's theorem and (1.74), we obtain that

$$I_2 = 2\int_t^{\hat{T}} \Big\langle \int_t^{\sigma} \Phi(\sigma,r)B(r)u(r)\mathrm{d}r,\ \Psi_2(\sigma)B(\sigma)u(\sigma)\Big\rangle \mathrm{d}\sigma.$$

Then by (1.75), (1.73) and (1.74), we find that

$$I_2 = 2\int_t^{\hat{T}} \Big\langle \int_t^{s} \Phi(s,r)B(r)u(r)\mathrm{d}r,\ \Psi_1(s)B(s)u(s)\Big\rangle \mathrm{d}s. \tag{1.78}$$

By (1.75), (1.73) and (1.74), using Fubini's theorem, we can easily verify that

$$I_4 = 2\int_t^{\hat{T}} \langle \Phi(s,t)h,\ \Psi_1(s)B(s)u(s)\rangle\, \mathrm{d}s.$$

Together with (1.78) and (1.70), the above yields that

$$I_2 + I_4 = \int_t^{\hat{T}} \langle y(s),\ \Psi_1(s)B(s)u(s)\rangle\, \mathrm{d}s. \tag{1.79}$$

Now from (1.71), (1.72), (1.76) and (1.79), we find that

$$\begin{aligned}
&\int_t^{\hat{T}} \big\| R^{1/2}\big(u(s) + R^{-1}B(s)^*\Upsilon(s)y(s;t,h,u)\big)\big\|^2\, \mathrm{d}s + \langle h,\ \Upsilon(t)h\rangle \\
&= \int_t^{\hat{T}} \langle u(s), Ru(s)\rangle \mathrm{d}s + \Big\langle h,\ \big[\Phi(\hat{T},t)^* M\Phi(\hat{T},t) + \int_t^{\hat{T}} \Phi(s,t)^* Q\Phi(s,t)\mathrm{d}s\big]h\Big\rangle \\
&\quad +2\int_t^{\hat{T}} \Big\langle y(s),\ \big[\Phi(\hat{T},s)^* M\Phi(\hat{T},s) + \int_s^{\hat{T}} \Phi(r,s)^* Q\Phi(r,s)\mathrm{d}r\big]B(s)u(s)\Big\rangle \mathrm{d}s.
\end{aligned} \tag{1.80}$$

By (1.52) and (1.80), we see that in order to show (1.69), it suffices to prove the following two identities:

$$\begin{aligned}
\langle y(\hat{T}), My(\hat{T})\rangle = &\Big\langle h,\ \Phi(\hat{T},t)^* M\Phi(\hat{T},t)h\Big\rangle \\
&+2\int_t^{\hat{T}} \Big\langle y(s),\ \Phi(\hat{T},s)^* M\Phi(\hat{T},s)B(s)u(s)\Big\rangle \mathrm{d}s
\end{aligned} \tag{1.81}$$

and

$$\begin{aligned}
\int_t^{\hat{T}} \langle y(s), Qy(s)\rangle \mathrm{d}s = &\Big\langle h,\ \int_t^{\hat{T}} \Phi(s,t)^* Q\Phi(s,t)\mathrm{d}s h\Big\rangle \\
&+2\int_t^{\hat{T}} \Big\langle y(s),\ \int_s^{\hat{T}} \Phi(r,s)^* Q\Phi(r,s)\mathrm{d}r\, B(s)u(s)\Big\rangle \mathrm{d}s.
\end{aligned} \tag{1.82}$$

We first show (1.81). By (1.70), we have that

$$\begin{aligned}
&\langle y(\hat{T}), My(\hat{T})\rangle \\
&= \Big\langle h,\ \Phi(\hat{T},t)^* M\Phi(\hat{T},t)h\Big\rangle + 2\Big\langle \Phi(\hat{T},t)h,\ M\int_t^{\hat{T}} \Phi(\hat{T},s)B(s)u(s)\mathrm{d}s\Big\rangle \\
&\quad +\Big\langle \int_t^{\hat{T}} \Phi(\hat{T},s)B(s)u(s)\mathrm{d}s,\ M\int_t^{\hat{T}} \Phi(\hat{T},s)B(s)u(s)\mathrm{d}s\Big\rangle.
\end{aligned} \tag{1.83}$$

Write J_1 for the last term on the right hand side of (1.83). By the similar reason to that showing (1.77), we can derive that

$$J_1 = 2\int_t^{\hat{T}} \Big\langle \int_t^s \Phi(s,r)B(r)u(r)\mathrm{d}r,\ \Phi(\hat{T},s)^* M\Phi(\hat{T},s)B(s)u(s)\Big\rangle \mathrm{d}s.$$

This, together with (1.70), indicates that

$$\begin{aligned} J_1 = 2\int_t^{\hat{T}} \Big\langle y(s),\ \Phi(\hat{T},s)^* M\Phi(\hat{T},s)B(s)u(s)\Big\rangle \mathrm{d}s \\ -2\Big\langle \Phi(\hat{T},t)h,\ \int_t^{\hat{T}} M\Phi(\hat{T},s)B(s)u(s)\mathrm{d}s\Big\rangle, \end{aligned}$$

which, along with (1.83), leads to (1.81). We next show (1.82). From (1.70), it follows that

$$\begin{aligned} &\int_t^{\hat{T}} \langle y(s), Qy(s)\rangle \mathrm{d}s \\ &= \Big\langle h, \int_t^{\hat{T}} \Phi(s,t)^* Q\Phi(s,t)\mathrm{d}s h\Big\rangle + 2\Big\langle h, \int_t^{\hat{T}} \Phi(s,t)^* Q\int_t^s \Phi(s,r)B(r)u(r)\mathrm{d}r\mathrm{d}s\Big\rangle \\ &+\int_t^{\hat{T}} \Big\langle \int_t^s \Phi(s,r)B(r)u(r)\mathrm{d}r,\ Q\int_t^s \Phi(s,\sigma)B(\sigma)u(\sigma)\mathrm{d}\sigma\Big\rangle \mathrm{d}s. \end{aligned} \tag{1.84}$$

Write J_2 for the third term on the right hand side of (1.84). By the similar reason to that showing (1.77), we can deduce that

$$J_2 = 2\int_t^{\hat{T}} \Big\langle \int_t^\sigma \Phi(\sigma,r)B(r)u(r)\mathrm{d}r,\ \int_\sigma^{\hat{T}} \Phi(\tau,\sigma)^* Q\Phi(\tau,\sigma)\mathrm{d}\tau B(\sigma)u(\sigma)\Big\rangle \mathrm{d}\sigma.$$

This, along with (1.70), as well as some computation, yields that

$$\begin{aligned} J_2 = 2\int_t^{\hat{T}} \Big\langle y(s),\ \int_s^{\hat{T}} \Phi(\tau,s)^* Q\Phi(\tau,s)\mathrm{d}\tau B(s)u(s)\Big\rangle \mathrm{d}s \\ -2\Big\langle h,\ \int_t^{\hat{T}} \Phi(\tau,t)^* \int_t^\tau Q\Phi(\tau,s)B(s)u(s)\mathrm{d}s\mathrm{d}\tau\Big\rangle. \end{aligned} \tag{1.85}$$

Now, (1.82) follows from (1.84) and (1.85). This completes the proof. □

The connection between $(LQ)_{t,h}^{\hat{T}}$ and Eq. (1.61) is stated in the following theorem.

Theorem 1.2 *Suppose that (1.53) holds. Let $(\bar{y}_{t,h}, \bar{u}_{t,h})$, with $t \in [0, \hat{T})$ and $h \in H$, be the optimal pair to Problem $(LQ)_{t,h}^{\hat{T}}$. Let Υ be the solution to Eq. (1.61). Then the following assertions hold: (i) The optimal trajectory $\bar{y}_{t,h}$ is the mild solution to the following closed loop equation:*

$$\begin{cases} y'(s) = A(s)y(s) - B(s)R^{-1}B(s)\Upsilon(s)y(s), & s \in [t, \hat{T}], \\ y(t) = h. \end{cases}$$

(ii) The optimal control $\bar{u}_{t,h}$ has the feedback form: $\bar{u}_{t,h}(\cdot) = -R^{-1}B(\cdot)^*\Upsilon(\cdot)\bar{y}_{t,h}(\cdot)$ over $[t, \hat{T}]$. (iii) The value function $W^{\hat{T}}$ verifies that $W^{\hat{T}}(t,h) = J^{\hat{T}}_{t,h}(\bar{u}_{t,h}) = \langle h, \Upsilon(t)h\rangle$, $t \in [0, \hat{T}]$, $h \in H$.

Proof Let $t \in [0, \hat{T}]$ and $h \in H$. It follows from the identity (1.69) that

$$J^{\hat{T}}_{t,h}(u) \geq \langle h, \Upsilon(t)h\rangle \quad \text{for each } u \in \mathscr{U}^{\hat{T}}_t. \tag{1.86}$$

Write $\hat{y}(\cdot)$ for the mild solution to the Eq. (1.2). Let $\hat{u}(\cdot) \triangleq -R^{-1}B(\cdot)^*\Upsilon(\cdot)\hat{y}(\cdot)$ over $[t, \hat{T}]$. We find that $\hat{y}(\cdot)$ and $y(\cdot; t, h, \hat{u})$ are the same. Setting $u = \hat{u}$ in (1.69) and taking into account (1.86), we see that $(\hat{y}, \hat{u})$ is an optimal pair to Problem $(LQ)^{\hat{T}}_{t,h}$. Then from the uniqueness of the optimal control to this problem, it follows that $\hat{y} = \bar{y}_{t,h}$ and $\hat{u} = \bar{u}_{t,h}$. Hence, (ii) and (iii) have been proved. From these, (i) follows at once. This ends the proof. □

1.2.2 Infinite Horizon Case

Given $t \geq 0$ and $h \in H$, we define a cost functional $J^{\infty}_{t,h} : L^2(t, \infty; U) \mapsto \mathbb{R}$ by setting

$$J^{\infty}_{t,h}(u) = \int_t^{\infty} \langle y(s; t, h, u), Qy(s; t, h, u)\rangle + \langle u(s), Ru(s)\rangle \mathrm{d}s, \quad u \in L^2(t, \infty; U). \tag{1.87}$$

(Recall that $y(\cdot; t, h, u)$ is the mild solution to Eq. (1.2).) Here and throughout this subsection, we assume that

$$Q \in S\mathscr{L}^+(H); \quad R \in S\mathscr{L}^+(U); \quad R \gg 0. \tag{1.88}$$

Notice that we only have $y(\cdot; t, h, u) \in L^2_{loc}(t, \infty; H)$, so $J^{\infty}_{t,h}(u)$ may take the value ∞. For each $t \geq 0$ and $h \in H$, we define the following LQ problem in the time horizon $[t, \infty)$:

$$(LQ)^{\infty}_{t,h} \qquad W^{\infty}(t,h) \triangleq \inf_{u \in L^2(t,\infty;U)} J^{\infty}_{t,h}(u). \tag{1.89}$$

Definition 1.10 (i) The map $(t,h) \to W^{\infty}(t,h)$, $(t,h) \in [0,\infty) \times H$, is called the value function associated with the family of LQ problems $\{(LQ)^{\infty}_{t,h}\}_{(t,h)\in[0,\infty)\times H}$. (ii) Given $t \geq 0$ and $h \in H$, $(LQ)^{\infty}_{t,h}$ satisfies the Finite Cost Condition (FCC, for short) if there is a control $u \in L^2(\mathbb{R}^+; U)$ so that $J^{\infty}_{t,h}(u) < \infty$. (iii) Given $t \geq 0$ and $h \in H$, $(LQ)^{\infty}_{t,h}$ is solvable if there is a control $\hat{u} \in L^2(\mathbb{R}^+; U)$ so that

$J^{\infty}_{t,h}(\hat{u}) = W^{\infty}(t,h)$. Such a control $\hat{u}$ is called an optimal control, $\hat{y}(\cdot) \triangleq y(\cdot; t, h, \hat{u})$ is called the corresponding optimal trajectory, while $(\hat{u}, \hat{y})$ is called an optimal pair to $(LQ)^{\infty}_{t,h}$.

Proposition 1.9 *(i) Let $t \in \mathbb{R}^+$ be given. If Problem $(LQ)^{\infty}_{t,h}$ satisfies the FCC for each $h \in H$, then there is a unique $\bar{\Upsilon}(t) \in S\mathscr{L}^+(H)$ so that*

$$W^{\infty}(t,h) = \langle h, \bar{\Upsilon}(t)h\rangle \ \ \textit{for each} \ \ h \in H. \tag{1.90}$$

(ii) If for all $t \geq 0$, $h \in H$, $(LQ)^{\infty}_{t,h}$ satisfies the FCC, then the map $t \to \bar{\Upsilon}(t)$ (where $\bar{\Upsilon}(s)$ is given by (1.90)) is T-periodic from $\mathbb{R}^+$ to $S\mathscr{L}^+(H)$. (iii) $(LQ)^{\infty}_{t,h}$ satisfies the FCC for all $(t,h) \in \mathbb{R}^+ \times H$ if and only if $(LQ)^{\infty}_{0,h}$ satisfies the FCC for all $h \in H$. (iv) If $(LQ)^{\infty}_{0,h}$ satisfies the FCC for all $h \in H$, then there is a unique T-periodic map $\bar{\Upsilon}(\cdot) : \mathbb{R}^+ \to S\mathscr{L}^+(H)$ verifying (1.90) for all $t \geq 0$.

Proof We begin with proving (i). Let $h \in H$. By the FCC of $(LQ)^{\infty}_{t,h}$ and the strict convexity of $J^{\infty}_{t,h}$, there is a unique optimal control $\bar{u}^{\infty}_{t,h}$ to $(LQ)^{\infty}_{t,h}$. Then we see that

$$W^k(t,h) \leq J^k_{t,h}\left(\bar{u}^{\infty}_{t,h}\big|_{[t,k]}\right) \leq W^{\infty}(t,h) \ \ \text{for each} \ \ k \in \mathbb{N},\ k > t, \tag{1.91}$$

where W^k is the value function to $\{(LQ)^k_{t,h}\}_{t\in[0,k],h\in H}$, with the same Q and R, and with $M = 0$. For any $k \in \mathbb{N}$ with $k > t$, let $\left(\bar{y}^k_{t,h}, \bar{u}^k_{t,h}\right)$ be the optimal pair of the above-mentioned problem $(LQ)^k_{t,h}$. Defined two functions $\widetilde{u}^k_{t,h}$ and $\widetilde{z}^k_{t,h}$ over $[t,\infty)$ by

$$\widetilde{u}^k_{t,h}(s) = \begin{cases} \bar{u}^k_{t,h}(s), & \text{when } s \in [t,k], \\ 0, & \text{when } s \in (k,\infty), \end{cases} \qquad \widetilde{z}^k_{t,h}(s) = \begin{cases} Q^{\frac{1}{2}}\bar{y}^k_{t,h}(s), & \text{when } s \in [t,k], \\ 0, & \text{when } s \in (k,\infty). \end{cases}$$

Notice that

$$\int_t^{\infty} \|\widetilde{z}^k_{t,h}(s)\|^2 \mathrm{d}s = \int_t^k \langle \bar{y}^k_{t,h}(s),\ Q\bar{y}^k_{t,h}(s)\rangle \mathrm{d}s \leq J^k_{t,h}(\bar{u}^k_{t,h}) = W^k(t,h)$$

and

$$\int_t^{\infty} \|\widetilde{u}^k_{t,h}(s)\|^2 \mathrm{d}s \leq \frac{1}{\delta}\int_t^k \langle \bar{u}^k_{t,h}(s),\ R\bar{u}^k_{t,h}(s)\rangle \mathrm{d}s \leq \frac{1}{\delta} J^k_{t,h}(\bar{u}^k_{t,h}) = \frac{1}{\delta} W^k(t,h),$$

where δ is a positive real number satisfying $R \geq \delta I$. (Since $R \gg 0$, such δ exist.) These, along with (1.91), yields that $\{\widetilde{u}^k_{t,h},\ k \in \mathbb{N},\ k > t\}$ and $\{\widetilde{z}^k_{t,h},\ k \in \mathbb{N},\ k > t\}$ are bounded in $L^2(t,\infty;U)$ and $L^2(t,\infty;H)$, respectively. Then, on subsequences, denoted in the same manner,

$$\widetilde{u}^k_{t,h} \to \hat{u} \ \text{ weakly in } \ L^2(t,\infty;U); \ \ \widetilde{z}^k_{t,h} \to \hat{z} \ \text{ weakly in } \ L^2(t,\infty;H), \tag{1.92}$$

as $k \to \infty$, for some $\hat{u}$ and $\hat{z}$. We now claim that

$$\hat{z}(s) = Q^{\frac{1}{2}} y(s; t, h, \hat{u}) \text{ for each } s > t. \tag{1.93}$$

Indeed, one can easily check that for each $s > t$,

$$y(\cdot; t, h; \widetilde{u}_{t,h}^k)\big|_{[t,s]} \to y(\cdot; t, h; \hat{u})\big|_{[t,s]} \text{ in } C([t, s]; L^2(\Omega)).$$

Thus, we find that

$$\widetilde{z}_{t,h}^k\big|_{[t,s]} \to Q^{\frac{1}{2}} y(\cdot; t, h; \hat{u})\big|_{[t,s]} \text{ in } C([t, s]; L^2(\Omega)).$$

This, along with the second condition of (1.92), indicates that

$$\hat{z}(\cdot)\big|_{[t,s]} = Q^{\frac{1}{2}} y(\cdot; t, h; \hat{u})\big|_{[t,s]} \text{ for each } s \geq t,$$

which leads to (1.93). Furthermore, it follows from (1.92) and (1.93) that

$$\begin{aligned} J_{t,h}^{\infty}(\hat{u}) &= \|\hat{z}\|_{L^2(t,\infty;H)}^2 + \left\| R^{\frac{1}{2}} \hat{u} \right\|_{L^2(t,\infty;U)}^2 \leq \lim_{k\to\infty} \left[\|\widetilde{z}_{t,h}^k\|_{L^2(t,\infty;H)}^2 + \|R^{\frac{1}{2}} \widetilde{u}_{t,h}^k\|_{L^2(t,\infty;U)}^2 \right] \\ &= \lim_{k\to\infty} J_{t,h}^k(\bar{u}_{t,h}^k) = \lim_{k\to\infty} W^k(t, h). \end{aligned}$$

Because $J_{t,h}^{\infty}(\hat{u}) \geq W^{\infty}(t, h)$, the above leads to $\lim_{k\to\infty} W^k(t, h) \geq W^{\infty}(t, h)$. This, together with (1.91), implies that

$$\lim_{k\to\infty} W^k(t, h) = W^{\infty}(t, h). \tag{1.94}$$

Given $k \in \mathbb{N}$ with $k > t$, write $(\bar{y}_{t,h}^k, \bar{u}_{t,h}^k)$, with $h \in H$, for the optimal pair to Problem $(LQ)_{t,h}^k$. By Theorem 1.2, one can easily verify that $(\bar{y}_{t,h}^k, \bar{u}_{t,h}^k)$ is linear with respect to $h \in H$. Thus we can define an linear operator $Q_k : H \to L^2(t, \infty; U) \times L^2(t, \infty; H)$ by $Q_k h = \left(R^{\frac{1}{2}} \widetilde{u}_{t,h}^k, \widetilde{z}_{t,h}^k \right)$, $h \in H$. Since

$$\left\| \left(R^{\frac{1}{2}} \widetilde{u}_h^k, \widetilde{z}_h^k \right) \right\|_{L^2(t,\infty;U)\times L^2(t,\infty;H)}^2 = J_{t,h}^k(\bar{u}_{t,h}^k)) = W^k(t, h),$$

we see that

$$\|Q_k h\|^2 = W^k(t, h) \text{ for each } h \in H. \tag{1.95}$$

This, along with (1.91), shows that $\|Q_k h\| \leq \sqrt{W^{\infty}(t, h)}$ for each $h \in H$. By the uniform boundedness theorem, there is $C > 0$ so that $\|Q_k\| \leq C$ for each $k \in \mathbb{N}$, with $k > t$. This, together with (1.94) and (1.95), yields that

$$W^{\infty}(t, h) = \lim_{k\to\infty} W^k(t, h) = \lim_{k\to\infty} \|Q_k h\|^2 \leq \|Q_k\|^2 \|h\|^2 \leq C^2 \|h\|^2. \tag{1.96}$$

We next show that $W^\infty(t,\cdot)$ satisfies the parallelogram law. Given $h_1, h_2 \in H$, let $\bar{u}^\infty_{t,h_1}$ and $\bar{u}^\infty_{t,h_2}$ be the optimal controls to $(LQ)^\infty_{t,h_1}$ and $(LQ)^\infty_{t,h_2}$, respectively. Then we have that

$$\begin{aligned}&W^\infty(t,h_1+h_2)+W^\infty(t,h_1-h_2)\\&\le J^\infty_{t,h_1+h_2}\left(\bar{u}^\infty_{t,h_1}+\bar{u}^\infty_{t,h_2}\right)+J^\infty_{t,h_1-h_2}\left(\bar{u}^\infty_{t,h_1}-\bar{u}^\infty_{t,h_2}\right)\\&=2\left(J^\infty_{t,h_1}(\bar{u}^\infty_{t,h_1})+J^\infty_{t,h_2}(\bar{u}^\infty_{t,h_2})\right)=2(W^\infty(t,h_1)+W^\infty(t,h_2)).\end{aligned} \tag{1.97}$$

Similarly, we can show that

$$W^\infty(t,g_1+g_2)+W^\infty(t,g_1-g_2)\le 2\left(W^\infty(t,g_1)+W^\infty(t,g_2)\right),$$

where $g_1=(h_1+h_2)/2$ and $g_2=(h_1-h_2)/2$. Because $W^\infty(t,ch)=c^2W^\infty(t,h)$ for all $c\in\mathbb{R}$ and $h\in H$, the above inequality implies that

$$2\left(W^\infty(t,h_1)+W^\infty(t,h_2)\right)\le W^\infty(t,h_1+h_2)+W^\infty(t,h_1-h_2).$$

This, together with (1.97), indicates that

$$W^\infty(t,h_1+h_2)+W^\infty(t,h_1-h_2)=2(W^\infty(t,h_1)+W^\infty(t,h_2)),$$

which leads to the desired parallelogram law. By this and (1.96), there is a unique $\bar{\Upsilon}(t)\in S\mathscr{L}^+(H)$ so that (1.90) holds (see [51]).

We now prove (ii). Since for each $t\ge 0$ and $h\in H$, Problem $(LQ)^\infty_{t,h}$ satisfies the FCC, it follows from (i) that $t\to\bar{\Upsilon}(t)$ defines a map from $\mathbb{R}^+\to S\mathscr{L}^+(H)$. Meanwhile, by the T-periodicity of $A(\cdot)$ and $B(\cdot)$, we find that for each $h\in H$, $W^\infty(\cdot,h)$ is T-periodic. This, along with (1.90), shows that $\bar{\Upsilon}(\cdot)$ is T-periodic.

We next show (iii). Assume that Problem $(LQ)^\infty_{0,h}$ satisfies the FCC for each $h\in H$. By the T-periodicity of $A(\cdot)$ and $B(\cdot)$, we find that for each $n\in\mathbb{N}$, $(LQ)^\infty_{nT,h}$ satisfies the FCC for all $h\in H$. Given $t\in[0,T)$, $h\in H$, write $\hat{h}\triangleq y(T;t,h,0)$. Because $(LQ)^\infty_{T,\hat{h}}$ satisfies the FCC, there is a control $u\in L^2(T,\infty;U)$ so that $J^\infty_{T,\hat{h}}(u)<\infty$. We now define $\hat{u}\in L^2(t,\infty;U)$ by

$$\hat{u}(s)=\begin{cases}0, & \text{when } s\in(t,T);\\ u(s), & \text{when } s\in[T,\infty).\end{cases}$$

Then we find that for each $t\in[0,T)$,

$$W^\infty(t,h)\le J^\infty_{t,h}(\hat{u})=\int_t^T\langle y(s;t,h,0),\ Qy(s;t,h,0)\rangle \mathrm{d}s+J^\infty_{T,\hat{h}}(u)<\infty.$$

Then, we use the FCC of $(LQ)^\infty_{nT,h}$ w.r.t. each $h\in H$ to repeat the above argument, with $n=2,\dots$, to get that $(LQ)^\infty_{t,h}$ also satisfies the FCC for each $t\ge 0$ and $h\in H$.

Finally, (iv) follows from (ii) and (iii). This ends the proof. □

Theorem 1.3 *Assume Problem $(LQ)^{\infty}_{0,h}$ satisfies the FCC for all $h \in H$. Let $\bar{\Upsilon}(t)$, with $t \geq 0$, be given by Proposition 1.9. Then the function $\bar{\Upsilon}(\cdot)$ is a mild solution to the following equation:*

$$\begin{cases} \dot{\Upsilon}(t) + A(t)^*\Upsilon(t) + \Upsilon(t)A(t) + Q - \Upsilon(t)B(t)R^{-1}B(t)^*\Upsilon(t) = 0, \quad t \in (0,T) \\ \Upsilon(T) = \Upsilon(0). \end{cases} \tag{1.98}$$

Proof First, we show the following Bellman optimality principle: When $t \in [0, T)$ and $h \in H$,

$$W^{\infty}(t,h) = \inf_{u \in L^2(t,+\infty;U)} \left\{ \int_t^T \Big[\langle y(s;t,h,u),\ Qy(s;t,h,u) \rangle + \langle u(s),\ Ru(s) \rangle \Big] ds + W^{\infty}(T, y(T;t,h,u)) \right\}. \tag{1.99}$$

Indeed, for any $u(\cdot) \in L^2((t,+\infty);U)$, let $u_1(\cdot) = u(\cdot)\big|_{[t,T]}$ and $u_2(\cdot) = u(\cdot)\big|_{(T,\infty)}$. Then

$$y(r;t,h,u) = y(r;T,y(T;t,h,u_1),u_2) \quad \text{for each } r \geq T. \tag{1.100}$$

Further, it follows from the definition of $W^{\infty}(t,h)$ and (1.100) that

$$\begin{aligned} &W^{\infty}(t,h) \\ &= \inf_{u_1 \in L^2(t,T;U)} \inf_{u_2 \in L^2(T,\infty;U)} \Bigg[\int_t^T [\langle y(s;t,h,u_1),\ Qy(s;t,h,u_1) \rangle + \langle u_1(s),\ Ru_1(s) \rangle] ds \\ &\quad + \int_T^{\infty} [\langle y(r;T,y(T;t,h,u_1),u_2),\ Qy(r;T,y(T;t,h,u_1),u_2) \rangle + \langle u_2(r),\ Ru_2(r) \rangle] dr \Bigg]. \end{aligned}$$

From this, we can easily verify (1.99).

Next, from (1.99), (1.90) and the periodicity of $\bar{\Upsilon}(\cdot)$, we see that when $t \in [0, T]$ and $h \in H$,

$$W^{\infty}(t,h) = \inf_{u(\cdot) \in L^2(t,T;U)} \left\{ \int_t^T [\langle y(s;t,h,u),\ Qy(s;t,h,u) \rangle + \langle u(s),\ Ru(s) \rangle]) ds + \langle y(T;t,h,u),\ \bar{\Upsilon}(0) y(T;t,h,u) \rangle \right\}. \tag{1.101}$$

From (1.101) and (iii) of Theorem 1.2, there is a function $\hat{\Upsilon} \in C_s([0,T]; S\mathscr{L}^+(H))$ solving the following equation in the mild sense:

$$\begin{cases} \dot{\Upsilon}(t) + A(t)^*\Upsilon(t) + \Upsilon(t)A(t) + Q - \Upsilon(t)B(t)R^{-1}B(t)^*\Upsilon(t),\ t \in (0,T) \\ \Upsilon(T) = \bar{\Upsilon}(0). \end{cases} \tag{1.102}$$

Moreover, we have that

$$W^{\infty}(t, h) = \langle h, \hat{\Upsilon}(t)h \rangle, \quad \text{when } t \in [0, T],\ h \in H. \tag{1.103}$$

This, along with (1.90), indicates that $\langle h, \hat{\Upsilon}(t)h \rangle = \langle h, \bar{\Upsilon}(t)h \rangle$ for any $t \in [0, T]$, $h \in H$. Because $\hat{\Upsilon}(t)$ and $\bar{\Upsilon}(t)$ are self-adjoint, we see that $\bar{\Upsilon}(\cdot)\big|_{[0,T]} = \hat{\Upsilon}(\cdot)$, which, together with (1.102), shows that $\bar{\Upsilon}$ is a mild solution of Eq. (1.98). This completes the proof. □

1.3 Relation Between Periodic Stabilization and LQ Problems

The aim of this section is to present some connection between the stabilization and LQ problems.

Definition 1.11 (i) Equation (1.1) is said to be linear periodic feedback stabilizable (LPFS, for short) if there is a T-periodic $K(\cdot) \in L^{\infty}\big(\mathbb{R}^+; \mathscr{L}(H, U)\big)$ so that the feedback equation:

$$y'(s) = \big[A + D(s) + B(s)K(s)\big]y(s),\ s \geq 0, \tag{1.104}$$

is exponentially stable, i.e., there are two positive constants C and δ so that for each $h \in H$,

$$\|y_K(s; 0, h)\| \leq Ce^{-\delta t}\|h\| \quad \text{for all } s \geq 0,\ h \in H, \tag{1.105}$$

where $y_K(\cdot; 0, h)$ denotes the solution of the Eq. (1.104) with the initial condition that $y(0) = h$. (ii) Any $K(\cdot)$ with the property in (i) is called an LPFS law for Eq. (1.1).

We now present the connection between the linear periodic feedback stablizability of Eq. (1.1) and LQ problems $(LQ)^{\infty}_{0,h}$. Notice that Problem $(LQ)^{\infty}_{0,h}$ depends on not only $h \in H$ but also Q and R.

Theorem 1.4 *The following assertions are equivalent:*

(i) Equation (1.1) is linear periodic feedback stablizable.
(ii) For any pair (Q, R) satisfying (1.88), the corresponding $(LQ)^{\infty}_{0,h}$ satisfies the FCC for each $h \in H$.
(iii) There is a pair (Q, R), with $Q \gg 0$ and $R \gg 0$, so that the corresponding $(LQ)^{\infty}_{0,h}$ satisfies the FCC for each $h \in H$.

Proof We first show that (i)⇒(ii). Suppose that Eq. (1.1) is linear periodic feedback stablizable. Then there is a T-periodic $K(\cdot) \in L^{\infty}\big(\mathbb{R}^+; \mathscr{L}(H, U)\big)$ so that (1.105) holds for some positive constants C and δ. Arbitrarily fix $h \in H$, construct a control u_h by setting

$$u_h(s) = K(s)y_K(s; 0, h), \quad s \geq 0. \tag{1.106}$$

It is clear that

$$y(s; 0, h, u_h) = y_K(s; 0, h), \quad s \geq 0. \tag{1.107}$$

Given (Q, R) satisfying (1.88), from (1.104), (1.88), (1.106) and (1.106), we find that

$$J_{0,h}^{\infty}(u) \leq \left(\|Q\| + \|R\| \|K\|^2_{L^\infty(\mathbb{R}^+; \mathscr{L}(H,U))}\right) \frac{C^2}{2\delta} \|h\|^2.$$

Hence, $(LQ)_{0,h}^{\infty}$ (corresponding to the pair (Q, R)) satisfies the FCC for any $h \in H$.

It is clear that (ii)$\Rightarrow$(iii), since (Q, R) clearly satisfies (1.88) when $Q \gg 0$ and $R \gg 0$.

We next show that (iii)$\Rightarrow$(i). Let $Q \gg 0$ and $R \gg 0$ so that the corresponding $(LQ)_{0,h}^{\infty}$ satisfies the FCC for all $h \in H$. By Theorem 1.3 and Lemma 1.9, there is a T-periodic operator-valued function $\bar{\Upsilon}(\cdot)$ (over $\mathbb{R}^+$), with (1.98) so that

$$W^{\infty}(t, h) = \langle h, \bar{\Upsilon}(t)h \rangle \quad \text{for any} \quad (t, h) \in [0, T] \times H. \tag{1.108}$$

Define $\bar{K}(\cdot) \in L^{\infty}(\mathbb{R}^+; \mathscr{L}(H, U))$ by setting

$$\bar{K}(s) = -R^{-1}B(s)^* \bar{\Upsilon}(s) \quad \text{for a.e.} \quad s \geq 0. \tag{1.109}$$

Clearly, $\bar{K}(\cdot)$ is T-periodic. We only need to show that the following equation is stable:

$$y'(s) = Ay(s) + D(s)y(s) + B(s)\bar{K}(s)y(s), \quad s \geq 0. \tag{1.110}$$

For this purpose, we write $\widetilde{y}(\cdot; 0, h)$ for the mild solution of Eq. (1.110) with the initial condition that $y(0) = h$, where $h \in H$ is arbitrarily given. For each $k \in \mathbb{N}$, define two functions:

$$\widetilde{y}_{k,h}(\cdot) \triangleq \widetilde{y}(\cdot; 0, h) \big|_{[0,kT]} \quad \text{and} \quad \widetilde{u}_{k,h}(\cdot) \triangleq \bar{K}(\cdot) \big|_{[0,kT]} \widetilde{y}_{k,h}(\cdot).$$

Clearly, they are in $C([0, kT]; H)$ and $L^2(0, kT; U)$ respectively. By Theorem 1.2, $(\widetilde{y}_{k,h}(\cdot), \widetilde{u}_{k,h}(\cdot))$ is the optimal pair to Problem $(LQ)_{0,h}^{kT}$ (corresponding to the above-mentioned (Q, R)). From the optimality of $(\widetilde{y}_{k,h}(\cdot), \widetilde{u}_{k,h}(\cdot))$, (iii) of Theorem 1.2 and the T-periodicity of $\bar{\Upsilon}(\cdot)$, we have

$$\begin{aligned} \langle h, \bar{\Upsilon}(0)h \rangle = \int_0^{kT} \big[\langle \widetilde{y}_{k,h}(s), \, Q\widetilde{y}_{k,h}(s) \rangle + \langle \widetilde{u}_{k,h}(s), \, R\widetilde{u}_{k,h}(s) \rangle\big] ds \\ + \langle \widetilde{y}_{k,h}(kT), \bar{\Upsilon}(0)\widetilde{y}_{k,h}(kT) \rangle. \end{aligned} \tag{1.111}$$

Since $Q \gg 0$, there is $q > 0$ so that $Q \geq qI$. Then, from (1.111), we find that

$$q \int_0^{kT} \|\widetilde{y}(s;0,h)\|^2 \mathrm{d}s \leq \int_0^{kT} \langle \widetilde{y}_{k,h}(s),\ Q\widetilde{y}_{k,h}(s)\rangle \mathrm{d}s \leq \langle h, \bar{\Upsilon}(0)h\rangle \text{ for all } k \in \mathbb{N} \text{ and } h \in H.$$

This leads to

$$\int_0^{\infty} \|\widetilde{y}(s;0,h)\|^2 \mathrm{d}s \leq \frac{1}{q}\|\bar{\Upsilon}(0)\|\|h\|^2 \text{ for each } h \in H. \tag{1.112}$$

Let $\Phi_{\widetilde{K}}$ be the evolution generated by $A(\cdot) + B(\cdot)\bar{K}(\cdot)$. Then $\sup_{r\in[0,T]} \|\Phi_{\widetilde{K}}(T,r)\| \leq C_1$ for some $C_1 > 0$. Fix an $h \in H$. We have that for each $s \in \mathbb{R}^+$,

$$\left\|\Phi_{\widetilde{K}}([s/T]T + T, 0)\, h\right\| \leq \left\|\Phi_{\widetilde{K}}(T, s - [s/T]T)\right\| \left\|\Phi_{\widetilde{K}}(s,0)\, h\right\| \leq C_1 \left\|\Phi_{\widetilde{K}}(s,0)\, h\right\|.$$

Hence, we find that

$$\left\|\Phi_{\widetilde{K}}(s,0)\, h\right\| \geq \left\|\Phi_{\widetilde{K}}([s/T]T + T, 0)\, h\right\| / C_1 \quad \text{for each } s \in \mathbb{R}^+.$$

This yields that

$$\left\|\Phi_{\widetilde{K}}(s,0)\, h\right\| \geq \left\|\Phi_{\widetilde{K}}(kT,0)\, h\right\| / C_1 \text{ for each } s \in [(k-1)T, kT) \text{ and } k \in \mathbb{N},$$

from which, it follows that

$$\int_{(k-1)T}^{kT} \left\|\Phi_{\widetilde{K}}(s,0)\, h\right\|^2 \mathrm{d}s \geq \frac{T}{C_1^2} \left\|\Phi_{\widetilde{K}}(kT,0)\, h\right\|^2 \text{ for each } k \in \mathbb{N}. \tag{1.113}$$

This, together with (1.112), indicates that for each $h \in H$,

$$\sum_{k=1}^{\infty} \|\Phi_{\widetilde{K}}(kT,0)h\|^2 \leq \frac{C_1^2}{qT}\|\bar{\Upsilon}(0)\|\|h\|^2,$$

from which, we find that for each $j \in \mathbb{N}$,

$$\|\Phi_{\widetilde{K}}(jT,0)\| \leq \sup_{h\in H\setminus\{0\}} \left\{\Big[\sum_{k=1}^{\infty} \|\Phi_{\widetilde{K}}(kT,0)h\|^2\Big]^{1/2} / \|h\|\right\} \leq \frac{C_1}{\sqrt{qT}}\sqrt{\|\bar{\Upsilon}(0)\|}. \tag{1.114}$$

Meanwhile, from (1.112), we see that function $\|\Phi_{\widetilde{K}}(\cdot,0)h\|^2 \in L^1(\mathbb{R}^+;\mathbb{R})$. This, together with (1.113), yields that

$$\lim_{k\to\infty} \Phi_{\widetilde{K}}(kT,0)h = 0 \text{ for any } h \in H. \tag{1.115}$$

Since $\Phi_{\tilde{K}}$ is compact, the following set is also compact:

$$\Phi_{\tilde{K}}(T,0)\mathscr{B}(0,1) = \{\Phi_{\tilde{K}}(T,0)h \mid h \in H,\ \|h\| \leq 1\}.$$

Then given $\varepsilon > 0$, there are $h_1,\ h_2,\ \ldots,\ h_{m_\varepsilon} \in H$ so that

$$\Phi_{\tilde{K}}(T,0)\mathscr{B}(0,1) \subseteq \bigcup_{j=1}^{m_\varepsilon} \mathscr{B}(h_j,\varepsilon).$$

Thus, for each $h \in \mathscr{B}(0,1)$, there is an index $j_h \in \{1,2,\ \ldots,\ m_\varepsilon\}$ so that

$$\|\Phi_{\tilde{K}}(T,0)h - h_{j_h}\| \leq \varepsilon. \tag{1.116}$$

By (1.115), there is $k_\varepsilon \in \mathbb{N}$ so that

$$\|\Phi_{\tilde{K}}(kT,0)h_j\| \leq \varepsilon \ \text{ for all } \ j \in \{1,2,\ldots,m_\varepsilon\},\ k \geq k_\varepsilon.$$

This, together with (1.116) and (1.114), implies that for each $h \in \mathscr{B}(0,1)$ and $k > k_\varepsilon$,

$$\|\Phi_{\tilde{K}}(kT,0)h\| \leq \left(1 + \frac{C_1}{\sqrt{qT}}\sqrt{\|\bar{\Upsilon}(0)\|}\right)\varepsilon.$$

Hence, $\lim_{k\to\infty} \|\Phi_{\tilde{K}}(kT,0)\| = 0$. Then there is $\hat{k} \in \mathbb{N}$ so that

$$\|\Phi_{\tilde{K}}^{\hat{k}}(T,0)\| = \|\Phi_{\tilde{K}}(\hat{k}T,0)\| \leq 1/2.$$

From this, we can easily verify that

$$\lim_{k\to\infty} \sqrt[k]{\|\Phi_{\tilde{K}}^k(T,0)\|} \leq \sqrt[\hat{k}]{1/2} < 1.$$

Then, by Lemma 1.1, Eq. (1.110) is exponentially stable. This completes the proof. □

Miscellaneous Notes

The LQ theory is an important field in the control theory. To our best knowledge, the first book introducing this subject is [8] (see Chap.4 of [8]). In 1960, the author of [43] introduced the relation between LQ problems and optimal feedback controls. The above-mentioned two monographs deal with finite dimensional systems. In the mid-1960s, the LQ theory was extended to partial differential equations in [61, 62]. The LQ theory for general evolution equations with bounded control operators was introduced in [66] and [26] in 1969 and 1976 respectively. The corresponding problems with unbounded control operators were studied in [4, 27, 62] respectively.

The LQ problems for parabolic, hyperbolic, and other partial differential equations with boundary controls were systematically studied in [54].

The stabilization is another important field of control theory. The fundamental theory of stability was established by A.M. Lyapunov in 1892 (see [70]). About the stabilization of differential equations, we would like to mention the following monographs: [36] (1975, it deals with some selected economic stabilization problems), [28] (1978, it introduces some sufficient conditions for exponential stabilizability connected with the decomposition of the spectrum of the infinitesimal generators), [12] (1987, it is devoted to robust stabilization of linear time invariant systems containing a real parameter vector), [53] (1989, it provides a comprehensive and unified treatment of uniform stabilization of the motion of a thin plate via boundary feedbacks), [48] (1994, it introduces the controllability and the stabilization of linear evolutionary systems of partial differential equations of conservative type), [50] (1998, it discusses the global stabilization and optimal control of nonlinear uncertain systems), [69] (1999, it studies the stability and the stabilization of some time invariant infinite dimensional systems), [24] (2007, it introduces different ways to design stabilizing feedback for both finite and infinite dimensional systems), [6] (2011, it studies the stabilization of Navier-Stokes flows), [83] (2012, it discusses the internal and external stabilization of linear time invariant finite dimensional systems with constraints). We also would like to mention the following papers which concern with the stabilization of time invariant finite dimensional systems: [1, 3, 5, 15, 16, 19–23, 25, 37, 39, 41, 42, 47, 49, 52, 57, 63–65, 80, 81, 84, 85, 87, 89, 91, 95, 96, 98]. About the periodic stabilization of periodic evolution equations, we here mention [7, 13, 31, 56, 68, 72, 82, 90, 94, 99].

The material in the first subsection is taken from [93]. The material in Sect. 1.2 is adapted from [29, 59]. For the last subsection, we do not find the exactly same results as those stated in Theorem 1.4. However, the similar results for autonomous systems can be found in [9–11], or [55]. Finally, we would like to explain the reason for us to introduce LQ problems in finite time horizon (in Sect. 1.2.1) as follows: It is well known that the stabilization is closely connected with LQ problems in infinite time horizon for autonomous systems. In this monograph, we introduce connection between the stabilization and LQ problems in finite time horizon for periodic systems and then study the stabilization for periodic systems with the aid of the aforementioned connection. These will be done in the next two chapters (see Theorem 3.1, for instance).

Chapter 2
Criteria on Periodic Stabilization in Infinite Dimensional Cases

Abstract This chapter studies the linear periodic feedback stabilization (LPFS, for short) for a class of evolution equations in the framework of Chap. 1. We restrict controls values only in a subspace Z of U, which might be of finite dimension.

Keywords Periodic Equations · Stabilization · Criteria · Infinite Dimension

Defnition 2.1 Equation (1.1) is said to be LPFS with respect to a subspace Z of U if there is a T-periodic $K(\cdot) \in L^\infty\big(\mathbb{R}^+; \mathscr{L}(H, Z)\big)$ so that the Eq. (1.104) is exponentially stable. Any such a $K(\cdot)$ is called an LPFS law for Eq. (1.1) with respect to Z. Write

$$\mathscr{U}^{FS} \triangleq \big\{Z \mid Z \text{ is a subspace of } U \text{ s.t. Eq. (1.1) is LPFS w.r.t. } Z\big\}. \tag{2.1}$$

We will provide three criteria for judging whether a subspace Z belongs to $\mathscr{U}^{FS}$. They are related with the following subjects: the attainable subspace of (1.1), which will be introduced in (2.2); the unstable subspace H_1 of (1.1) with the null control, which was defined in (1.21); the periodic map associated to (1.1) with the null control, which was defined in (1.12); and two unique continuation properties for the dual equations of (1.1) on different time horizons $[0, T]$ and $[0, n_0 T]$ (where n_0 was defined by (1.17)), which will be introduced in (2.57) and (2.58) respectively. We also show that if $U \in \mathscr{U}^{FS}$, then there is a finite dimensional subspace Z in $\mathscr{U}^{FS}$. Hence, Eq. (1.1) is LPFS if and only if it is LPFS with respect to a finite dimensional subspace Z of U. This might help us to design feedback laws numerically.

2.1 Attainable Subspaces

This section is devoted to studies of attainable subspaces w.r.t. a subspace $Z \subset U$. Let

G. Wang and Y. Xu, *Periodic Feedback Stabilization for Linear Periodic Evolution Equations*, SpringerBriefs in Mathematics,
DOI 10.1007/978-3-319-49238-4_2

$$\mathscr{A}_k^Z \triangleq \left\{ \int_0^{kT} \Phi(kT,s)B(s)u(s)\mathrm{d}s \;\middle|\; u(\cdot) \in L^2(\mathbb{R}^+; Z) \right\} \quad \text{for all} \quad k \in \mathbb{N}. \tag{2.2}$$

It is called the kT-attainable subspace of Eq. (1.1) w.r.t. Z. Recall (1.12), (1.19) and (1.20). We simply write

$$H_1 \triangleq H_1(0), \;\; H_2 \triangleq H_2(0), \;\; \mathbb{P} \triangleq \mathbb{P}(0) \;\; \text{and} \;\; \mathscr{P} \triangleq \mathscr{P}(0). \tag{2.3}$$

Let

$$\hat{\mathscr{A}}_k^Z = \mathbb{P}\mathscr{A}_k^Z, \qquad k \in \mathbb{N}. \tag{2.4}$$

These subspaces play important roles in our studies of LPFS.

Lemma 2.1 *Let $[A(\cdot), B(\cdot)]$ be a T-periodic pair satisfying the conditions $(\mathscr{H}_1)$-$(\mathscr{H}_2)$. Assume that Q and R satisfy (1.88) and $Q \gg 0$. Let $h \in H$. Then Problem $(LQ)_{0,h}^{\infty}$ (defined by (1.54) with $\hat{T} = T$, $M = 0$ and $U = Z$) satisfies the FCC at h if*

$$\mathbb{P}\big(\mathscr{P}^k h\big) \in \hat{\mathscr{A}}_k^Z \;\; \textit{for some} \;\; k \in \{0, 1, 2, \dots\}. \tag{2.5}$$

Proof Suppose that $h \in H$ satisfies (2.5) for some $k \in \mathbb{N}$. Then there is $u \in L^2(0, \infty; Z)$ so that

$$\mathbb{P}\big(\mathscr{P}^k h\big) = \mathbb{P} \int_0^{kT} \Phi(kT,s)B(s)u(s)\mathrm{d}s,$$

from which, it follows that $\mathbb{P}y(kT; 0, h, -u) = 0$. Let $\hat{u} = -\chi_{[0,kT]}u$. Then $\hat{u}$ and $y(kT; 0, h, \hat{u})$ are in $L^2(t, \infty; Z)$ and H_2 respectively. These, together with (f) of Proposition 1.4, yield that

$$\begin{aligned} &\|y(s; 0, h, \hat{u})\| = \|y(s; kT, y(kT; 0, h, \hat{u}), \hat{u})\| \\ &= \|y(s; kT, y(kT; 0, h, \hat{u}), 0)\| = \|\Phi(s, kT)y(kT; 0, h, \hat{u})\| \\ &\le C_{\bar{\rho}} e^{-\rho(s-kT)} \|y(kT; 0, h, \hat{u})\| \;\; \text{for all} \;\; s \ge kT. \end{aligned}$$

From this, one can easily verify that $J_{0,h}^{\infty}(\hat{u}) < \infty$. So Problem $(LQ)_{0,h}^{\infty}$ satisfies the FCC at h.

We next suppose that $h \in H$ satisfies (2.5) with $k = 0$. Since h satisfies (2.5), with $k = 0$, if and only if $h \in H_2$, we find that $J_{0,h}^{\infty}(0) < \infty$. So Problem $(LQ)_{0,h}^{\infty}$ satisfies the FCC at h. This ends the proof. □

From Theorem 1.4 and Lemma 2.1, we find that properties of subspaces $\{\mathscr{A}_k^Z, k \in \mathbb{N}\}$ play important roles in the studies of LPFS. Meanwhile, since h satisfies (2.5), with $k = 0$, if and only if $h \in H_2$, we find from Lemma 2.1 that when $h \in H_2$, Problem $(LQ)_{0,h}^{\infty}$ satisfies the FCC at h. Hence, the studies of the case when $h \in H_1$ is very important. We will see that it isindeed the key in the studies of LPFS.

Since H_1 is invariant with respect to $\mathscr{P}$ (see (*b*) of Proposition 1.4), we can define $\mathscr{P}_1 : H_1 \to H_1$ by setting

$$\mathscr{P}_1 \triangleq \mathscr{P}\big|_{H_1}. \tag{2.6}$$

Then, by (1.22), it follows that

$$\sigma(\mathscr{P}_1) \bigcap \mathbb{B} = \varnothing. \tag{2.7}$$

Lemma 2.2 *Let $\mathscr{P}_1$ and n_0 be given by (2.6) and (1.17), respectively. Suppose that $Z \subseteq U$ is a subspace with $\mathscr{A}_k^Z$ and $\hat{\mathscr{A}}_k^Z$ given by (2.2) and (2.4), respectively. Then for each $k \in \mathbb{N}$,*

$$\mathscr{A}_k^Z = \mathscr{A}_1^Z + \mathscr{P}\mathscr{A}_1^Z + \cdots + \mathscr{P}^{k-1}\mathscr{A}_1^Z; \; \hat{\mathscr{A}}_k^Z = \hat{\mathscr{A}}_1^Z + \mathscr{P}_1\hat{\mathscr{A}}_1^Z + \cdots + \mathscr{P}_1^{k-1}\hat{\mathscr{A}}_1^Z. \tag{2.8}$$

Furthermore, $\mathscr{P}_1$ is invertible and

$$\hat{\mathscr{A}}^Z = \hat{\mathscr{A}}_{n_0}^Z; \quad \mathscr{P}_1\hat{\mathscr{A}}^Z = \hat{\mathscr{A}}^Z = \mathscr{P}_1^{-1}\hat{\mathscr{A}}^Z, \tag{2.9}$$

where

$$\hat{\mathscr{A}}^Z \triangleq \bigcup_{k=1}^{\infty} \hat{\mathscr{A}}_k^Z. \tag{2.10}$$

Proof We begin with proving the first equality in (2.8) by the mathematical induction. Clearly, it stands when $k = 1$. Assume that it holds in the case when $k = k_0$ for some $k_0 \geq 1$, i.e.,

$$\mathscr{A}_{k_0}^Z = \mathscr{A}_1^Z + \mathscr{P}\mathscr{A}_1^Z + \cdots + \mathscr{P}^{k_0-1}\mathscr{A}_1^Z. \tag{2.11}$$

Because of (1.23) and (2.3), we have that $\Phi((k_0+1)T, T) = \Phi(T, 0)^{k_0} = \mathscr{P}^{k_0}$. This, along with (2.2), the T-periodicity of $B(\cdot)$ and (2.11), indicates that

$$\begin{aligned}
\mathscr{A}_{k_0+1}^Z &= \Big\{ \int_0^{(k_0+1)T} \Phi((k_0+1)T, s)B(s)u(s)\mathrm{d}s \;\Big|\; u(\cdot) \in L^2(\mathbb{R}^+; Z) \Big\} \\
&= \mathscr{P}^{k_0}\mathscr{A}_1^Z + \Big\{ \int_0^{k_0 T} \Phi(k_0 T, s)B(s)u(s+T)\mathrm{d}s \;\Big|\; u(\cdot) \in L^2(\mathbb{R}^+; Z) \Big\} \\
&= \mathscr{P}^{k_0}\mathscr{A}_1^Z + \mathscr{A}_{k_0}^Z = \mathscr{A}_1^Z + \mathscr{P}\mathscr{A}_1^Z + \cdots + \mathscr{P}^{k_0}\mathscr{A}_1^Z.
\end{aligned}$$

which leads to the first equality in (2.8).

We next show the second equality in (2.8). By (2.3) and (1.23), we have that $\mathscr{P}\mathbb{P} = \mathbb{P}\mathscr{P}$. Since $\mathbb{P}$ is a projection from H onto H_1 (see Proposition 1.4), the above, along with the first equality in (2.8) and (2.4), leads to the second equality in (2.8).

Then we show the first equality in (2.9). It follows from (2.10) and (2.8) that

$$\hat{\mathscr{A}}_{n_0}^Z \subseteq \hat{\mathscr{A}}^Z \;\text{ and }\; \hat{\mathscr{A}}_k^Z \subseteq \hat{\mathscr{A}}_{n_0}^Z, \;\text{ when } k \leq n_0. \tag{2.12}$$

Since $\dim H_1 = n_0$ (see (1.22)) and $\mathscr{P}_1 : H_1 \to H_1$ (see (2.6)), according to the Hamilton-Cayley theorem, each $\mathscr{P}_1^j$ with $j \geq n_0$ is a linear combination of $\{ I, \mathscr{P}_1^1, \mathscr{P}_1^2, \cdots, \mathscr{P}_1^{(n_0-1)}\}$. This, along with the second equality in (2.8), indicates that

$$\hat{\mathscr{A}}_k^Z = \sum_{j=0}^{k-1} \mathscr{P}_1^j(\hat{\mathscr{A}}_1^Z) \subseteq \sum_{j=0}^{n_0-1} \mathscr{P}_1^j(\hat{\mathscr{A}}_1^Z) = \hat{\mathscr{A}}_{n_0}^Z, \quad \text{when } k \geq n_0. \tag{2.13}$$

Now the first equality in (2.9) follows from (2.12) and (2.13).

Finally, we show the non-singularity of $\mathscr{P}_1$ and the second equality in (2.9). By the first equality in (2.9) and the Hamilton-Cayley theorem, we see that

$$\mathscr{P}_1\hat{\mathscr{A}}^Z = \mathscr{P}_1\hat{\mathscr{A}}_{n_0}^Z = \mathscr{P}_1 \sum_{j=0}^{n_0-1} \mathscr{P}_1^j(\hat{\mathscr{A}}_1^Z) = \sum_{j=1}^{n_0} \mathscr{P}_1^j(\hat{\mathscr{A}}_1^Z) \subseteq \sum_{j=0}^{n_0-1} \mathscr{P}_1^j(\hat{\mathscr{A}}_1^Z) = \hat{\mathscr{A}}_{n_0}^Z,$$

from which, it follows that

$$\mathscr{P}_1\hat{\mathscr{A}}^Z \subseteq \hat{\mathscr{A}}^Z. \tag{2.14}$$

From (1.22) and (1.22), we find that $0 \notin \sigma(\mathscr{P}_1^C)$ and H_1^C (the domain of $\mathscr{P}_1^C$) is of finite dimension. Thus $\mathscr{P}_1^C$ is invertible. So is $\mathscr{P}_1$. This implies that $\dim(\mathscr{P}_1\hat{\mathscr{A}}^Z) = \dim\hat{\mathscr{A}}^Z$, which, together with (2.14), yields that $\mathscr{P}_1\hat{\mathscr{A}}^Z = \hat{\mathscr{A}}^Z$. This completes the proof. □

2.2 Three Criterions on Periodic Feedback Stabilization

The aim of this section is to prove the following theorem.

Theorem 2.1 *Let $\mathbb{P}$, $\mathscr{P}$ and H_j with $j = 1, 2$ be given by (2.3). Let n_0 be given by (1.17). Then, for each subspace $Z \subseteq U$, the following assertions are equivalent:*

(a) Equation (1.1) is LPFS with respect to Z, i.e., $Z \in \mathscr{U}^{FS}$.
(b) $\hat{\mathscr{A}}_{n_0}^Z = H_1$, where $\hat{\mathscr{A}}_{n_0}^Z$ is given by (2.4).
(c) If $\xi \in \mathbb{P}^* H_1$ and $\left(B(\cdot)\big|_Z\right)^* \Phi(n_0T, \cdot)^*\xi = 0$ over $(0, n_0T)$, then $\xi = 0$.
(d) If $\xi \in H^C$ satisfies that $(\mu I - \mathscr{P}^{*C})\xi = 0$, with $\mu \notin \mathbb{B}$, and $\left(B(\cdot)\big|_Z\right)^{*C} \Phi(T, \cdot)^{*C} \xi = 0$ over $(0, T)$, then $\xi = 0$.

2.2.1 Multi-periodic Feedback Stabilization

In this subsection, we will introduce three propositions. The first two propositions will be used in the proof of Theorem 2.1. The last one is independently interesting.

Defnition 2.2 (i) Equation (1.1) is said to be linear multi-periodic feedback stabilizable (LMPFS, for short) if there is a kT-periodic $K(\cdot) \in L^\infty\left(\mathbb{R}^+; \mathscr{L}(H, U)\right)$ for some $k \in \mathbb{N}$ so that Eq. (1.104) is exponentially stable. Any such a $K(\cdot)$ is called an LMPFS law for Eq. (1.1).
(ii) Equation (1.1) is said to be LMPFS with respect to a subspace Z of U if there is a kT-periodic $K(\cdot) \in L^\infty\left(\mathbb{R}^+; \mathscr{L}(H, Z)\right)$ for some $k \in \mathbb{N}$ so that Eq. (1.104) is exponentially stable. Any such a $K(\cdot)$ is called an LMPFS law for Eq. (1.1) with respect to Z.

Proposition 2.1 *Let n_0 and H_1 be given by (1.17) and (2.3) respectively. Suppose that $Z \subseteq U$ satisfies (b) of Theorem 2.1. Then, Eq. (1.1) is LMPFS with respect to Z.*

Proof Let $Z \subseteq U$ satisfy (b) of Theorem 2.1. We organize the proof by several steps as follows.

Step 1. To construct, for each $h_1 \in H_1$, $u^{h_1}(\cdot) \in L^2(\mathbb{R}^+; Z)$ so that $\mathbb{P}y(n_0T; 0, h_1, u^{h_1}) = 0$.

Because $\dim H_1 = n_0$ (see (1.22)), we can set $\{\eta_1, \cdots, \eta_{\hat{n}}\}$ to be an orthonormal basis of H_1. By (1.21), we see that $\mathbb{P}\mathscr{P}^{n_0}\eta_j \in H_1$ for each $j \in \{1, 2, \ldots, n_0\}$. Then it follows from (b) of Theorem 2.1 that $\mathbb{P}\mathscr{P}^{n_0}\eta_j \in \hat{\mathscr{A}}^Z_{n_0}$. Thus for each j, there is $u_j \in L^2(\mathbb{R}^+; Z)$ so that

$$\mathbb{P}\mathscr{P}^{n_0}\eta_j = \mathbb{P}\int_0^{n_0T} \Phi(n_0T, s)B(s)\hat{u}_j(s)\mathrm{d}s. \tag{2.15}$$

For any $h_1 \in H_1$, let

$$u^{h_1} = -\chi_{(0,n_0T)}\sum_{j=1}^{n_0}\langle h_1, \eta_j\rangle u_j. \tag{2.16}$$

Then we define an operator $\mathscr{L} : H_1 \to L^2(\mathbb{R}^+; Z)$ by setting

$$\mathscr{L}h_1(\cdot) = u^{h_1}(\cdot) \;\text{ for all }\; h_1 \in H_1. \tag{2.17}$$

Clearly, $\mathscr{L}$ is linear and bounded. Since $h_1 = \sum_{j=1}^{n_0}\langle h_1, \eta_j\rangle\eta_j$, it follows from (2.16) and (2.15) that

$$\begin{aligned}
&\mathbb{P}y(n_0T; 0, h_1, \mathscr{L}h_1)\\
&= \sum_{j=1}^{n_0}\langle h_1, \eta_j\rangle\mathbb{P}\mathscr{P}^{n_0}\eta_j - \sum_{j=1}^{n_0}\langle h_1, \eta_j\rangle\mathbb{P}\int_0^{n_0T}\Phi(n_0T, s)B(s)u_j(s)\mathrm{d}s \quad (2.18)\\
&= \sum_{j=1}^{n_0}\langle h_1, \eta_j\rangle\left(\mathbb{P}\mathscr{P}^{n_0}\eta_j - \mathbb{P}\int_0^{n_0T}\Phi(n_0T, s)B(s)u_j(s)\mathrm{d}s\right) = 0.
\end{aligned}$$

Step 2. To show the existence of an $N_0 \in \mathbb{N}$ so that

$$\|y(NT;0,h,\mathscr{L}(\mathbb{P}h))\| \le \delta_0\|h\| \quad \textit{for all} \ \ h\in H \ \ \textit{and} \ \ N\ge N_0, \tag{2.19}$$

where

$$\delta_0 \triangleq (1+\hat{\delta})/2<1, \quad \textit{with} \ \ \hat{\delta} \ \ \textit{given by} \ \ (1.16) \tag{2.20}$$

Let $\rho_0 = -\ln\delta_0/T$. Then $0<\rho_0<-\ln\hat{\delta}/T \triangleq \hat{\rho}$. By (f) of Proposition 1.4, there is a constant $C_{\rho_0}>0$ so that

$$\|y(kT;0,h_2,0)\| = \|\Phi(kT,0)h_2\| \le C_{\rho_0}e^{-\rho_0 kT}\|h_2\| = C_{\rho_0}\delta_0^k\|h_2\|, \tag{2.21}$$

when $k\in\mathbb{N}$ and $h_2\in H_2$. We claim that there is a constant $C>0$ so that

$$\|y(NT;0,h_1,\mathscr{L}h_1)\| \le CC_{\rho_0}\delta_0^{N-n_0}\|h_1\|, \ \ \text{when} \ \ h_1\in H_1, N\ge n_0, \tag{2.22}$$

where $\mathscr{L}$ is given by (2.17). In fact, because for each $h_1\in H_1$,

$$\left\|y(n_0T;0,h_1,\mathscr{L}h_1)\right\| \le \left\|\Phi(n_0T,0)h_1\right\| + \left\|\int_0^{n_0T}\Phi(n_0T,s)B(s)\mathscr{L}h_1(s)\mathrm{d}s\right\|,$$

we see that there is a constant $C>0$ so that

$$\|y(n_0T;0,h_1,\mathscr{L}h_1)\| \le C\|h_1\| \ \ \text{for all} \ \ h_1\in H_1. \tag{2.23}$$

Here, we used facts that $B(\cdot)\in L^\infty(\mathbb{R}^+;\mathscr{L}(U;H))$ and $\mathscr{L}$ is linear and bounded. Meanwhile, it follows from (2.16) that $\mathscr{L}h_1(\cdot)=0$ over $(n_0T,+\infty)$. This, together with (1.8) (where $u(\cdot)\equiv 0$), yields that for all $N\ge n_0$ and $h_1\in H$,

$$\begin{aligned}
&y(NT;0,h_1,\mathscr{L}h_1) = y(NT;n_0T,y(n_0T;0,h_1,\mathscr{L}h_1),\mathscr{L}h_1)\\
&= y(NT;n_0T,y(n_0T;0,h_1,\mathscr{L}h_1),0) = \Phi(NT,n_0T)y(n_0T;0,h_1,\mathscr{L}h_1)\\
&= \Phi((N-n_0)T,0)y(n_0T;0,h_1,\mathscr{L}h_1) = y((N-n_0)T;0,y(n_0T;0,h_1,\mathscr{L}h_1),0).
\end{aligned} \tag{2.24}$$

Because of (2.18) and (1.20)–(1.21) with $t=0$, we see that $y(n_0T;0,h_1,\mathscr{L}h_1)\in H_2$, when $h_1\in H_1$. This along with (2.24), (2.21) and (2.23), leads to (2.22).

Let

$$N_0 = \max\left\{\frac{\ln C_{\rho_0} + \ln\left(C\delta_0^{-n_0}\|\mathbb{P}\| + \|I-\mathbb{P}\|\right)}{\ln(1/\delta_0)} + 2, \ n_0\right\}. \tag{2.25}$$

(Here, $[r]$ with $r\in\mathbb{R}$ denotes the integer so that $r-1<[r]\le r$.) Then, it follows from (2.22), (2.21) and (2.25) that for all $N\ge N_0$, $h\in H$,

$$\|y(NT;0,h,\mathscr{L}(\mathbb{P}h))\| \le C_{\rho_0}\delta_0^N(C\delta_0^{-n_0}\|\mathbb{P}\| + \|I-\mathbb{P}\|)\|h\| \le \delta_0\|h\|.$$

Step 3. To study the value function associated with a family of LQ problems

Given $N \in \mathbb{N}$, $t \in [0, NT)$ and $h \in H$, write $y_N^Z(\cdot; t, h, u) \in C([t, NT]; H)$ for the solution to the equation:

$$\begin{cases} y'(s) = Ay(s) + D(s)y(s) + B(s)\big|_Z u(s) \text{ in } (t, NT), \\ y(t) = h, \end{cases} \tag{2.26}$$

where $B(s)\big|_Z$ is the restriction of $B(s)$ on the subspace Z. For each $\varepsilon > 0$, define the cost functional:

$$J_{t,h}^{NT,\varepsilon,Z}(u) = \int_t^{NT} \varepsilon \|u(s)\|_U^2 ds + \|y_N^Z(NT; t, h, u)\|^2, \quad u \in L^2(t, NT; Z). \tag{2.27}$$

Then consider the following LQ problem

$$(LQ)_{t,h}^{NT,\varepsilon,Z}: \qquad W^{NT,\varepsilon,Z}(t, h) \triangleq \inf_{u \in L^2(t,NT;Z)} J_{t,h}^{NT,\varepsilon,Z}(u).$$

Let

$$\varepsilon_0 \stackrel{\Delta}{=} (\delta_0 - \delta_0^2)/(\|\mathscr{L}\|\|P\| + 1)^2, \text{ with } \delta_0 \text{ and } \mathscr{L} \text{ given by (2.20) and (2.17).} \tag{2.28}$$

We claim that when N_0 is given by (2.25),

$$W^{NT,\varepsilon,Z}(0, h) \le \delta_0 \|h\|^2 \text{ for all } h \in H, \text{ when } N \ge N_0 \text{ and } \varepsilon \in (0, \varepsilon_0]. \tag{2.29}$$

In fact, it follows from (2.28) that for each $h \in H$ and $\varepsilon \in (0, \varepsilon_0]$,

$$\varepsilon \|\mathscr{L}(\mathbb{P}h)(\cdot)\|_{L^2(\mathbb{R}^+;Z)}^2 \le \varepsilon_0 \|\mathscr{L}\|^2 \|\mathbb{P}\|^2 \|h\|^2 \le (\delta_0 - \delta_0^2)\|h\|^2. \tag{2.30}$$

Since $\mathbb{P}$ is a projection from H to H_1 (see Proposition 1.4), it follows from (2.17) that

$$\mathscr{L}(\mathbb{P}h) \in L^2(\mathbb{R}^+; Z) \text{ for all } h \in H. \tag{2.31}$$

Since

$$y_N^Z\left(\cdot; t, h, u\big|_{[t,NT)}\right) = y(\cdot; t, h, u)\big|_{[t,NT]} \text{ for any } u \in L^2(t, +\infty; Z),$$

we see that

$$y_N^Z\left(NT; 0, h, \mathscr{L}(\mathbb{P}h)\big|_{(0,NT)}\right) = y(NT; 0, h, \mathscr{L}(\mathbb{P}h)).$$

This, together with (2.27), (2.31), (2.30) and (2.19), indicates that

$$W^{NT,\varepsilon,Z}(0,h) \leq \varepsilon \int_0^{NT} \|\mathscr{L}(Ph)(s)\|^2 ds + \|y_N^Z(NT;0,h,\mathscr{L}(\mathbb{P}h))\|^2 \leq \delta_0 \|h\|^2,$$

when $N \geq N_0$, $\varepsilon \in (0,\varepsilon_0]$, $h \in H$, i.e., (2.29) stands.

Step 4. To construct an NT-periodic $K^Z_{\varepsilon,N}(\cdot) \in L^\infty(\mathbb{R}^+;\mathscr{L}(H,Z))$.

Arbitrarily fix an $\varepsilon \in (0,\varepsilon_0]$ and an $N \geq N_0$, where N_0 and ε_0 are given by (2.25) and (2.28) respectively. By Theorem 1.2, we can derive that

$$W^{NT,\varepsilon,Z}(t,h) = \langle Q^{NT,\varepsilon,Z}(t)h,h\rangle \quad \text{for all} \quad h \in H,$$

where $Q^{NT,\varepsilon,Z}(\cdot)$ is the solution to the Riccati equation:

$$\begin{cases} \dot{\Upsilon}(t) + A(t)^*\Upsilon(t) + \Upsilon(t)A(t) - \frac{1}{\varepsilon}\Upsilon(t)B(t)\big|_Z \big(B(t)\big|_Z\big)^* \Upsilon(t) = 0, & t \in (0,NT), \\ \Upsilon(NT) = I. \end{cases} \tag{2.32}$$

Meanwhile, it follows form (2.27) that for each $h \in H$,

$$0 \leq \left\langle h, Q^{NT,\varepsilon,Z}(t)h \right\rangle \leq J^{NT,\varepsilon,Z}_{t,h}(0) \leq \|\Phi(NT,t)\|^2 \|h\|^2.$$

Define $K^Z_{\varepsilon,N}(\cdot) : [0,NT) \to \mathscr{L}(H;Z)$ by

$$K^Z_{\varepsilon,N}(t) = -\frac{1}{\varepsilon}\left(B(s)\big|_Z\right)^* Q^{NT,\varepsilon,Z}(t) \quad \text{for a.e.} \quad t \in [0,NT). \tag{2.33}$$

One can easily check that $K^Z_{\varepsilon,N}(\cdot) \in L^\infty(0,NT;\mathscr{L}(H;Z))$. From this and $(\mathscr{H}_1)$-$(\mathscr{H}_2)$, the following feedback equation has a unique mild solution $y^Z_{\varepsilon,N}(\cdot;0,h) \in C([0,NT];H)$:

$$\begin{cases} y'(s) = Ay(s) + D(s)y(s) + B(s)\big|_Z K^Z_{\varepsilon,N}(s)y(s) \quad \text{in} \quad (0,NT), \\ y(0) = h \in H. \end{cases} \tag{2.34}$$

Define the following function:

$$u^Z_{\varepsilon,N,0,h}(s) \triangleq K^Z_{\varepsilon,N}(s)y^Z_{\varepsilon,N}(s;0,h) \qquad \text{for a.e.} \quad s \in (0,NT). \tag{2.35}$$

By Theorem 1.2, we see that $u^Z_{\varepsilon,N,0,h}(\cdot)$ is the optimal control to $(LQ)^{NT,\varepsilon,Z}_{0,h}$. This yields that

$$W^{NT,\varepsilon,Z}(0,h) = J^{NT,\varepsilon,Z}_{0,h}\left(u^Z_{\varepsilon,N,0,h}\right), \quad \text{when} \quad h \in H. \tag{2.36}$$

By (2.26) with $t = 0$, (2.34) and (2.35), we see that

$$y^Z_{\varepsilon,N}(NT;0,h) = y^Z_N(NT;0,h,u^Z_{\varepsilon,N,0,h}).$$

From this, (2.27), (2.36) and (2.29), it follows that

$$\|y^Z_{\varepsilon,N}(NT;0,h)\|^2 \le J^{NT,\varepsilon,Z}_{0,h}(u^Z_{\varepsilon,N,0,h}) = W^{NT,\varepsilon,Z}(0,h) \le \delta_0\|h\|^2, \quad \text{when } h \in H. \tag{2.37}$$

Now, we extend NT-periodically $K^Z_{\varepsilon,N}(\cdot)$ over $\mathbb{R}^+$ by setting

$$K^Z_{\varepsilon,N}(t+kNT) = K^Z_{\varepsilon,N}(t) \ \text{ for all } \ t \in [0,NT),\ k \in \mathbb{N}. \tag{2.38}$$

Step 5. To prove that when $\varepsilon \in (0,\varepsilon_0]$ and $N \ge N_0$, $K^Z_{\varepsilon,N}(\cdot)$ *is an LMPFS law for Eq.* (1.1) *with respect to* Z.

Consider the feedback equation:

$$\begin{cases} y'(s) = Ay(s) + D(s)y(s) + B(s)\big|_Z K^Z_{\varepsilon,N}(s)y(s) & \text{in } \mathbb{R}^+, \\ y(0) = h \in H. \end{cases} \tag{2.39}$$

Since $K^Z_{\varepsilon,N}(\cdot) \in L^\infty(\mathbb{R}^+;\mathscr{L}(H;Z))$, we have that $D(\cdot)+B(\cdot)\big|_Z K^Z_{\varepsilon,N}(\cdot)$ belongs to $L^1_{loc}(\mathbb{R}^+;\mathscr{L}(H))$ (see $(\mathscr{H}_1)$ and $(\mathscr{H}_2)$). Thus, for each $h \in H$, Eq. (2.39) has a unique mild solution $y^Z_\varepsilon(\cdot;0,h)$. Clearly, we have that $y^Z_\varepsilon(t;0,h) = y^Z_{\varepsilon,N}(t;0,h)$ for each $t \in [0,NT]$. Write $\Phi^Z_{\varepsilon,N}$ for the evolution generated by $A(\cdot)+B(\cdot)\big|_Z K^Z_{\varepsilon,N}(\cdot)$. By Proposition 1.2, we find that

$$y^Z_{\varepsilon,N}(NT;0,h) = \Phi^Z_{\varepsilon,N}(NT,0)h \quad \text{for all } \ h \in H.$$

This, along with (2.37) and (2.20), yields that

$$\|\Phi^Z_{\varepsilon,N}(NT,0)\| \le \sqrt{\delta_0} < 1. \tag{2.40}$$

Since $D(\cdot)$ and $B(\cdot)$ are T-periodic and $K^Z_{\varepsilon,N}(\cdot)$ is NT-periodic, it follows that

$$\Phi^Z_{\varepsilon,N}(s+NT,t+NT) = \Phi^Z_{\varepsilon,N}(s,t), \quad \text{when } \ 0 \le t \le s < +\infty. \tag{2.41}$$

By (2.41) and (2.40), we see that Eq. (2.39) is exponentially stable. Hence, $K^Z_{\varepsilon,N}(\cdot)$, with $\varepsilon \in (0,\varepsilon_0]$ and $N \ge N_0$, is an LMPFS law for Eq. (2.39). This ends the proof. □

Proposition 2.2 *Let Z be a subspace of U. Then, Eq. (1.1) is LPFS with respect to Z if and only if it is LMPFS with respect to Z.*

Proof Clearly, Eq. (1.1) is LPFS w.r.t. Z $\Rightarrow$ Eq. (1.1) is LMPFS w.r.t. Z. To show the reverse, we suppose that Eq. (1.1) is LMPFS with respect to Z. Then there is an NT-periodic $\hat{K}^Z_N(\cdot) \in L^\infty(\mathbb{R}^+;\mathscr{L}(H;Z))$, with $N \in \mathbb{N}$, so that the following feedback equation is exponentially stable:

$$y'(s) = Ay(s) + D(s)y(s) + B(s)\big|_Z \hat{K}^Z_N(s)y(s), \quad s \ge 0. \tag{2.42}$$

Because $\hat{K}_N^Z(\cdot) \in L^\infty(\mathbb{R}^+; \mathscr{L}(H; Z))$ is NT-periodic, from the above, as well as assumptions $(\mathscr{H}_1)$-$(\mathscr{H}_2)$, one can verify that for each $t \geq 0$ and $h \in H$, the solution $\hat{y}_N^Z(\cdot; t, h)$ to the equation:

$$\begin{cases} y'(s) + Ay(s) + D(s)y(s) = B(s)\big|_Z \hat{K}_N^Z(s)y(s) \text{ in } [t, \infty), \\ y(t) = h \end{cases} \tag{2.43}$$

satisfies that

$$\|\hat{y}_N^Z(s; t, h)\| \leq C_1 e^{-\delta_1(s-t)}\|h\|, \text{ when } s \geq t \text{ and } h \in H, \tag{2.44}$$

where C_1 and δ_1 are two positive constants independent of h, t and s. Given $h \in H$, let $u^h(\cdot) \triangleq \hat{K}_N^Z(\cdot)\hat{y}_N^Z(\cdot; 0, h)$. It follows from (2.44) that $u^h(\cdot) \in L^2(\mathbb{R}^+; Z)$. Moreover, it holds that $y(\cdot; 0, h, u^x) = \hat{y}_N^Z(\cdot; 0, h)$. Consider the LQ problem $(LQ)_{0,h}^\infty$, defined by (1.89) where $U = Z$, $Q = I_H$ and $R = I_U$. It follows from (2.44) that for each $h \in H$,

$$\int_t^\infty \big[\|u^h(s)\|^2 + \|y(s; 0, h, u^h)\|^2\big]ds \leq \int_t^\infty \big[(\|\hat{K}_N^Z(s)\|^2 + 1)\|y(s; 0, h, u^h)\|^2\big]ds$$
$$\leq \big(\|\hat{K}_N^Z\|^2 + 1\big)\int_t^\infty \|\hat{y}_N^Z(s; t, h)\|^2 ds \leq \frac{M_1^2}{2\delta_1}\big(\|\hat{K}_N^Z\|^2 + 1\big)\|h\|^2.$$

Therefore, Problem $(LQ)_{0,h}^\infty$ satisfies the FCC for any $h \in H$. Then by Theorem 1.4, we see that Eq. (1.1) is LMPFS with respect to Z. This ends the proof. □

Proposition 2.3 *When both $D(\cdot)$ and $B(\cdot)$ are time invariant, i.e., $D(t) \equiv D$ and $B(t) \equiv B$ for all $t \geq 0$, the following statements are equivalent:*
(a) Equation (1.1) is linear $\hat{T}$-periodic feedback stabilizable for some $\hat{T} > 0$.
(b) Equation (1.1) is linear $\hat{T}$-periodic feedback stabilizable for any $\hat{T} > 0$.
(c) Equation (1.1) is linear time invariant feedback stabilizable.

Proof It suffices to show that $(a) \Rightarrow (c)$. Let $N \in \mathbb{N}$ with $N \geq 2$ and let $T = \hat{T}/N$. Since $D(\cdot)$ and $B(\cdot)$ are time invariant, Eq. (1.1) is T-periodic. Because of (a), there is an NT-periodic $\hat{K}_N^U(\cdot) \in L^\infty(\mathbb{R}^+; \mathscr{L}(H; U))$ so that the feedback Eq. (2.42), where $Z = U$, is exponentially stable. By Theorem 1.4, there is a pair (Q, R), with $Q \gg 0$ and $R \gg 0$, so that the Problem $(LQ)_{0,h}^\infty$ satisfies the FCC for each $h \in H$. Now, by the same way to show $(iii) \Rightarrow (i)$ in the proof of Theorem 1.4 (where $Z = U$), we see that $\bar{K}(\cdot)$, given by (1.109) with $Z = U$, is a LPFS law for Eq. (1.1). We claim that this $\bar{K}(\cdot)$ is time invariant in the case that $D(\cdot)$ and $B(\cdot)$ are time invariant. When this is done, $\bar{K}(\cdot) \equiv \bar{K} \in \mathscr{L}(H; U)$ is a feedback law for Eq. (1.1), which leads to (c).

The rest is to show that $\bar{K}(\cdot)$ is time invariant. Let $W^\infty(t, h)$ be the value function given by (1.89) with the aforementioned Q and R. By the time invariance of $B(\cdot)$, and by (1.108) and (1.109), it suffices to show $W^\infty(t, h)$ is time invariant. The later will be proved as follows. Since Eq. (1.1) is time invariant, it follows that when $t \in \mathbb{R}^+$, $h \in H$ and $u(\cdot) \in L^2(\mathbb{R}^+; U)$, we have that $y(s; t, h, u) = y(s - t; 0, h, v)$ for all

$s \geq t$, where $v(\cdot)$ is defined by $v(s) = u(s+t)$ for all $s \geq 0$. Hence, given $t \in \mathbb{R}^+$ and $h \in H$, we have that for each $u(\cdot) \in L^2(\mathbb{R}^+; U)$,

$$\int_t^\infty \left(\|Q^{1/2}y(s;t,h,u(s))\|^2 + \|R^{1/2}u(s)\|_U^2\right) \mathrm{d}s$$
$$= \int_0^\infty \left(\|Q^{1/2}y(r;0,h,u(r+t))\|^2 + \|R^{1/2}u(r+t)\|_U^2\right) \mathrm{d}r.$$

Taking the infimum on the both sides of the above equation with respect to $u(\cdot) \in L^2(\mathbb{R}^+; U)$ leads to $W^\infty(t,h) = W^\infty(0,h)$. So the value function $W^\infty(t,h)$ is independent of t. This completes the proof. $\square$

Remark 2.1 By Proposition 2.3, we see that linear time-periodic functions $K(\cdot)$ will not aid in the linear stabilization of Eq. (1.1) when both $D(\cdot)$ and $B(\cdot)$ are time invariant. On the other hand, when Eq. (1.1) is T-periodically time varying, linear time-periodic functions $K(\cdot)$ do aid in the linear stabilization of Eq. (1.1). This can be seen from the following 2-periodic ODE: $y'(t) = \sum_{j=1}^{\infty} \left[\chi_{[2j,2j+1)}(t) - \chi_{[2j+1,2j+2)}(t)\right] u(t)$, $t \geq 0$. For each $k \in \mathbb{R}$, consider the equation: $y'(t) = \sum_{j=1}^{\infty} \left[\chi_{[2j,2j+1)}(t) - \chi_{[2j+1,2j+2)}(t)\right] ky(t)$, $t \geq 0$. Clearly, the corresponding periodic map $\mathscr{P}_k \equiv 1$. Thus any linear time invariant feedback equation is not exponentially stable. On the other hand, by a direct computation, one can easily check that the following 2-periodic map is an LPFS law:

$$k(t) = \sum_{j=1}^{\infty} \left[\chi_{[2j,2j+1)}(t) + 2\chi_{[2j+1,2j+2)}(t)\right], \quad t \geq 0.$$

2.2.2 *Proof of Theorem 2.1*

We first show that $(a) \Leftrightarrow (b)$. To prove that $(b) \Rightarrow (a)$, suppose that $Z \subseteqq U$ satisfies (b) in Theorem 2.1. By Proposition 2.1, we see that Eq. (1.1) is LMPFS with respect to Z. This, along with Proposition 2.2, yields (a).

To verify that $(a) \Rightarrow (b)$, we suppose, by contradiction, that $Z \in \mathscr{U}^{FS}$, but (b) in Theorem 2.1 did not hold. Then $\hat{\mathscr{A}}_{n_0}^Z$ would be a proper subspace of H_1. This, along with (2.9), yields that $\hat{V}^Z$ is a proper subspace of H_1. One can directly check that

$$\left(H_1 \bigcap (\hat{\mathscr{A}^Z})^\perp\right) \perp \hat{\mathscr{A}^Z}; \quad H_1 = \hat{\mathscr{A}^Z} \bigoplus \left(H_1 \bigcap (\hat{\mathscr{A}^Z})^\perp\right). \tag{2.45}$$

Since $\hat{\mathscr{A}^Z}$ is a proper subspace of H_1 and $\dim H_1 = n_0$ (see (1.22)), we have that

$$n_0 \geq l \triangleq \dim \left(H_1 \bigcap (\hat{\mathscr{A}^Z})^\perp\right) \geq 1. \tag{2.46}$$

By (2.45) and (2.46), we can let $\{\eta_1, \ldots, \eta_{n_0}\}$ be a basis of H_1 so that $\{\eta_1, \cdots, \eta_l\}$ and $\{\eta_{l+1}, \cdots, \eta_{n_0}\}$ are bases of $H_1 \bigcap (\hat{\mathscr{A}}^Z)^\perp$ and $\hat{\mathscr{A}}^Z$. By (2.9) in Lemma 2.2, $\hat{\mathscr{A}}^Z$ is an invariant subspace under $\mathscr{P}_1$. Thus there are matrices $A_1 \in \mathbb{R}^{l\times l}$, $A_2 \in \mathbb{R}^{(n_0-l)\times l}$, $A_3 \in \mathbb{R}^{(n_0-l)\times(n_0-l)}$ so that

$$\mathscr{P}_1\left(\eta_1, \cdots, \eta_{n_0}\right) = \left(\eta_1, \cdots, \eta_{n_0}\right)\begin{pmatrix} A_1 & 0_{l\times(n_0-l)} \\ A_2 & A_3 \end{pmatrix}. \tag{2.47}$$

Let P_{11} be the orthogonal projection from H_1 onto $H_1 \bigcap (\hat{\mathscr{A}}^Z)^\perp$. Define a linear bijection $\mathscr{J} : \mathbb{R}^l \to \left(H_1 \bigcap (\hat{\mathscr{A}}^Z)^\perp\right)$ by setting

$$\mathscr{J}(\alpha) \triangleq (\eta_1, \ldots, \eta_l)\alpha \quad \text{for each column vector } \alpha \in \mathbb{R}^l. \tag{2.48}$$

By (2.47) and (2.48), we see that for all $\alpha \in \mathbb{R}^l$ and $k \in \mathbb{N}$,

$$\begin{aligned} & P_{11}\mathscr{P}_1^k\mathscr{J}(\alpha) = P_{11}\mathscr{P}_1^k(\eta_1, \cdots, \eta_l)\alpha \\ & = P_{11}(\eta_1, \cdots, \eta_l, \mid \eta_{l+1}, \cdots, \eta_{n_0})\begin{pmatrix} A_1 & 0_{l\times(n_0-l)} \\ A_2 & A_3 \end{pmatrix}^k \begin{pmatrix} \alpha \\ 0_{(n_0-l)\times 1} \end{pmatrix} \\ & = (\eta_1, \cdots, \eta_l)A_1^k\alpha. \end{aligned} \tag{2.49}$$

On the other hand, since $Z \in \mathscr{U}^{FS}$, there is a T-periodic $K(\cdot) \in L^\infty(\mathbb{R}^+; \mathscr{L}(H; Z))$ so that Eq. (1.104) is exponentially stable, which implies that for all $h \in H$,

$$\lim_{t\to+\infty} y_K(t; 0, h) = 0, \tag{2.50}$$

where $y_K(\cdot; 0, h)$ denotes the solution of Eq. (1.104) with the initial condition that $y(0) = h$. Let $u_K^h(t) \triangleq K(t)y_K(t; 0, h)$ for a.e. $t \geq 0$. Then by (1.8), we have that

$$y_K(t; 0, h) = \Phi(t, 0)h + \int_0^t \Phi(t, s)B(s)u_K^h(s)\mathrm{d}s, \quad \text{when } t \in \mathbb{R}^+ \text{ and } h \in H. \tag{2.51}$$

From (2.51) and (2.2), it follows that

$$P_{11}Py_K(kT; 0, h) \in P_{11}\mathbb{P}(\mathscr{P}^k h + \mathscr{A}_k^Z) \quad \text{for all } h \in H \text{ and } k \in \mathbb{N}. \tag{2.52}$$

Since $\mathbb{P}\mathscr{A}_k^Z \triangleq \hat{\mathscr{A}}_k^Z \subseteq \hat{\mathscr{A}}^Z$ for all $k \in \mathbb{N}$ (see (2.4) and (2.10)) and P_{11} is the orthogonal projection from H_1 onto $H_1 \bigcap (\hat{\mathscr{A}}^Z)^\perp$, it follows from (2.45) that $P_{11}P\mathscr{A}_k^Z \subset P_{11}\hat{\mathscr{A}}^Z = \{0\}$. Because $\mathbb{P}\mathscr{P}^k = \mathscr{P}^k\mathbb{P}$ for all $k \in \mathbb{N}$ (see Parts (a) and (e) in Proposition 1.4), the above, along with (2.52), indicates that

$$P_{11}\mathbb{P}y_K(kT; 0, h) = P_{11}\mathbb{P}\mathscr{P}^k h = P_{11}\mathscr{P}^k\mathbb{P}h \quad \text{for all } h \in H \text{ and } k \in \mathbb{N}. \tag{2.53}$$

Since $\mathbb{P} : H \to H_1$ is a projection (see Proposition 1.4), from (2.50) and (2.53), we see that

$$\lim_{k\to+\infty} P_{11}\mathscr{P}^k h = 0, \quad \text{when } h \in H_1. \tag{2.54}$$

Now by (2.54), (2.49) and (2.6), we have that $\lim_{k\to\infty} A_1^k\alpha = 0$, when $\alpha \in \mathbb{R}^l$. Thus, we have that

$$\sigma(A_1) \in \mathbb{B} \ \ \text{(the open unit ball in } \mathbb{C}^1\text{)}. \tag{2.55}$$

By (2.47), we find that $\sigma(A_1) \subset \sigma(\mathscr{P}_1)$. This, together with (2.55) and (2.7), leads to a contradiction. Hence, $(a) \Rightarrow (b)$. This completes the proof of $(a) \Leftrightarrow (b)$.

We next show that $(b) \Leftrightarrow (c)$. First of all, we mention that (b) means that Eq. (1.1) over $(0, n_0T)$ is null controllable under the projection $\mathbb{P}$ with respect to the initial data $h \in H_1$, while (c) says that the adjoint equation

$$\psi_t(t) + A^*\psi(t) + D(t)^*\psi(t) = 0 \ \text{ for a.e. } t \in (0, n_0T), \quad \psi(n_0T) = \xi \tag{2.56}$$

with the initial data in $\mathbb{P}^* H_1$ has the unique continuation property. Such two properties are equivalent in finite dimensional spaces. The detailed proof is as follows. We introduce two complex adjoint equations as follows:

$$\psi'(t) + A^{*C}\psi(t) + D(t)^{*C}\psi(t) = 0 \ \text{ in } \ (0, n_0T), \quad \psi(n_0T) \in H^C; \tag{2.57}$$

$$\psi'(t) + A^{*C}\psi(t) + D(t)^{*C}\psi(t) = 0 \ \text{ in } \ (0, T), \quad \psi(T) \in H^C. \tag{2.58}$$

For each $\xi \in H^C$, Eq. (2.57) (or (2.58)) with the initial condition that $\psi^{\xi}_{n_0}(n_0T) = \xi$ (or $\psi^{\xi}(T) = \xi$) has a unique solution in $C([0, n_0T]; H^C)$ (or $C([0, T]; H^C)$). We denote this solution by $\psi^{\xi}_{n_0}(\cdot)$ (or $\psi^{\xi}(\cdot)$). Clearly, when $\xi \in H$, $\psi^{\xi}_{n_0}(\cdot) \in C([0, n_0T]; H)$ and $\psi^{\xi}(\cdot) \in C([0, T]; H)$ are accordingly the solutions of (2.57) and (2.58) where A^C and $D(t)^C$ are replaced by A and $D(t)$ respectively. One can easily check that

$$\psi^{\xi}(0) = \mathscr{P}^{*C}\xi \ \text{ and } \ \psi^{\xi}_{n_0}(0) = \left(\mathscr{P}^{*C}\right)^{n_0}\xi \ \text{ for all } \ \xi \in H^C. \tag{2.59}$$

By the T-periodicity of $D^*(\cdot)$, we see that for each $\xi \in H^C$,

$$\psi^{\xi}_{n_0}((k-1)T+t) = \psi^{\xi_k}(t), \ t \in [0, T], \ k \in \{1, \dots, n_0\}, \ \text{ where } \xi_k \triangleq \left(\mathscr{P}^{*C}\right)^{n_0-k}\xi. \tag{2.60}$$

Now we carry out the proof of $(b) \Leftrightarrow (c)$ by several steps as follows.

Step 1. To prove that (b) is equivalent to the following property:

$$\forall\, h \in H, \ \exists\, u^h(\cdot) \in L^2(\mathbb{R}^+; Z) \ s.t. \ \mathbb{P}y(n_0T; 0, h, u^h) = 0 \tag{2.61}$$

Suppose that (b) holds. Then by (2.2), we have that

$$\mathbb{P}\Big\{\int_0^{n_0T}\Phi(n_0T,t)B(t)u(t)\mathrm{d}t \;\Big|\; u(\cdot)\in L^2(\mathbb{R}^+;Z)\Big\}=H_1. \tag{2.62}$$

Given $h\in H$, it holds that $\mathbb{P}\Phi(n_0T,0)h\in H_1$ (see (1.21)). From this and (2.62), there is a $u^h(\cdot)\in L^2(\mathbb{R}^+;Z)$ so that

$$\mathbb{P}y(n_0T;0,h,u^h)=\mathbb{P}\Phi(n_0T,0)h+\mathbb{P}\int_0^{n_0T}\Phi(n_0T,t)B(t)u^h(t)\mathrm{d}t=0,$$

which leads to (2.61).

Conversely, assume that (2.61) holds. Then for any $h\in H$, there exists $u^h(\cdot)\in L^2(\mathbb{R}^+;Z)$ so that $\mathbb{P}y(n_0T;0,h,u^h)=0$. Thus, we find that

$$\begin{aligned}H_1\supseteq\hat{\mathscr{A}}^Z_{n_0}&\triangleq\mathbb{P}\Big\{\int_0^{n_0T}\Phi(n_0T,t)B(t)u(t)\mathrm{d}t\;\Big|\;u(\cdot)\in L^2(\mathbb{R}^+;Z)\Big\}\\&\supseteq\mathbb{P}\Big\{\int_0^{n_0T}\Phi(n_0T,t)B(t)u^h(t)\mathrm{d}t\;\Big|\;h\in H\Big\}\\&=-\mathbb{P}\Big\{\Phi(n_0T,0)h\;\Big|\;h\in H\Big\}=\mathbb{P}\mathscr{P}^{n_0}H.\end{aligned} \tag{2.63}$$

Since $\mathbb{P}\mathscr{P}=\mathscr{P}\mathbb{P}$ (see (1.23)), $\mathbb{P}H=H_1$ and $\mathscr{P}H_1=\mathscr{P}_1H_1=H_1$ (see (2.6) and Lemma 2.2), we see that $\mathbb{P}\mathscr{P}^{n_0}H=H_1$. This, together with (2.63), leads to (b).

Step 2. To show that $\xi\in\mathbb{P}^*H_1$ *and* $\psi^{\xi}_{n_0}(0)=0\Rightarrow\xi=0$

Recall Proposition 1.5. Because $\tilde{H}_1$ is invariant under $\mathscr{P}^*$, it follows from (2.59) that

$$\psi^{\xi}_{n_0}(0)=(\mathscr{P}^*)^{n_0}\xi=\big(\mathscr{P}^*\big|_{\tilde{H}_1}\big)^{n_0}\xi\in\tilde{H}_1,\ \text{ when }\xi\in\tilde{H}_1. \tag{2.64}$$

By Proposition 1.5, we find that $\sigma\big(\mathscr{P}^{*C}|_{\tilde{H}_1^C}\big)\bigcap\mathscr{B}=\varnothing$ and $\dim\tilde{H}_1=n_0<\infty$. Thus, the map $\big(\mathscr{P}^*\big|_{\tilde{H}_1}\big)^{n_0}$ is invertible from $\tilde{H}_1$ onto $\tilde{H}_1$. Then by (2.64), we see that $\xi=0$, when $\xi\in\tilde{H}_1$ and $\psi^{\xi}_{n_0}(0)=0$. This, together with (1.43), implies that $\xi=0$, when $\xi\in P^*H_1$ and $\psi^{\xi}_{n_0}(0)=0$.

Step 3. To show that (2.61) $\Rightarrow$ (c)

Clearly, when η, $h\in H$ and $u(\cdot)\in L^2(\mathbb{R}^+;Z)$, we have that

$$\langle\psi^{\eta}_{n_0}(0),\,h\rangle=\langle\eta,\,y(n_0T;0,h,u)\rangle-\int_0^{n_0T}\langle(B(t)|_Z)^*\psi^{\eta}_{n_0}(t),\,u(t)\rangle\mathrm{d}t. \tag{2.65}$$

Suppose that ξ satisfies conditions in (c). Then by (2.65) where $\eta=\xi$ and $\psi^{\xi}_{n_0}(t)=\Phi(n_0T,t)^*\xi$, we find that

$$\langle\psi^{\xi}_{n_0}(0),\,h\rangle=\langle\xi,\,y(n_0T;0,h,u)\rangle,\ \text{ when }h\in H\text{ and }u(\cdot)\in L^2(\mathbb{R}^+;Z). \tag{2.66}$$

By (2.61), given $h \in H$, there is a $u^h(\cdot) \in L^2(\mathbb{R}^+; Z)$ so that

$$\mathbb{P}y(n_0T; 0, h, u^h) = 0. \tag{2.67}$$

Since $\xi \in \mathbb{P}^* H_1$, there is $g \in H_1$ with $\xi = \mathbb{P}^* g$. This, along with (2.66) and (2.67), indicates that

$$\begin{aligned}\langle \psi_{n_0}^{\xi}(0),\ h\rangle &= \langle \xi,\ y(n_0T; 0, h, u^h)\rangle = \langle \mathbb{P}^* g,\ y(n_0T; 0, h, u^h)\rangle\\ &= \langle g,\ \mathbb{P}y(n_0T; 0, h, u^h)\rangle = 0 \ \text{ for all } \ h \in H.\end{aligned}$$

Hence, $\psi_{n_0}^{\xi}(0) = 0$. Then by the conclusion of Step 2, we have that $\xi = 0$. So (c) holds.

Step 4. To show that $(c) \Rightarrow$ (2.61)

Assume that (c) holds. Define two subspaces

$$\begin{aligned}\Gamma &\triangleq \left\{ \left(B(\cdot)\big|_Z\right)^* \psi_{n_0}^{\xi}(\cdot) \mid \xi \in \mathbb{P}^* H_1 \right\} \subseteq L^2(0, n_0T; Z);\\ \Gamma_0 &\triangleq \left\{ \psi_{n_0}^{\xi}(0) \mid \xi \in \mathbb{P}^* H_1 \right\} \subseteq H.\end{aligned} \tag{2.68}$$

By (c) and the conclusion of Step 2, we see that the following map $\mathscr{L}_1 : \Gamma \rightarrow \Gamma_0$ is well defined:

$$\mathscr{L}_1 \left(\left(B(\cdot)\big|_Z\right)^* \psi_{n_0}^{\xi}(\cdot) \right) = \psi_{n_0}^{\xi}(0) \quad \text{for all} \quad \xi \in P^* H_1. \tag{2.69}$$

Clearly, it is linear. Given $h \in H$, define a linear functional $\mathscr{F}^h$ on Γ by

$$\mathscr{F}^h(\gamma) = \langle \mathscr{L}_1(\gamma), h\rangle \quad \text{for all} \quad \gamma \in \Gamma. \tag{2.70}$$

Since $\dim(\mathbb{P}^* H_1) = \dim \tilde{H}_1 = n_0 < \infty$, it holds that $\dim \Gamma < \infty$. Thus, $\mathscr{F}^h \in \mathscr{L}(\Gamma; \mathbb{R})$. By the Hahn-Banach theorem , there is a $\bar{\mathscr{F}}^h \in \mathscr{L}(L^2(0, n_0T; Z); \mathbb{R})$ so that

$$\bar{\mathscr{F}}^h(\gamma) = \mathscr{F}^h(\gamma) \ \text{ for all } \ \gamma \in \Gamma; \ \text{ and } \ \|\bar{\mathscr{F}}^h\| = \|\mathscr{F}^h\|. \tag{2.71}$$

Then by the Riesz representation theorem (see p. 59 in [32]), there exists a function $u^h(\cdot)$ in $L^2(0, n_0T; Z)$ so that

$$\bar{\mathscr{F}}^h(\gamma) = -\int_0^{n_0T} \langle u^h(t), \gamma(t)\rangle_U \mathrm{d}t \ \text{ for all } \ \gamma \in L^2(0, n_0T; Z). \tag{2.72}$$

Since $\mathbb{P}^* H_1 = \mathbb{P}^* H$ (see (1.42)), it follows from (2.69) to (2.72) that

$$-\int_0^{n_0T} \langle \left(B(t)\big|_Z\right)^* \psi_{n_0}^{\mathbb{P}^*\hat{\eta}}(t),\ u^h(t)\rangle \mathrm{d}t = \langle \psi_{n_0}^{\mathbb{P}^*\hat{\eta}}(0),\ h\rangle \ \text{ for all } \ \hat{\eta} \in H.$$

Meanwhile, it follows by (2.65) that for each $\hat{\eta} \in H$,

$$\langle \psi_{n_0}^{\mathbb{P}^*\hat{\eta}}(0), h \rangle = \langle \mathbb{P}^*\hat{\eta}, y(n_0T; 0, h, u^h) \rangle - \int_0^{n_0T} \langle (B(t)|_Z)^* \psi_{n_0}^{\mathbb{P}^*\hat{\eta}}(t), u^h(t) \rangle dt.$$

The above two equalities imply that

$$\langle \hat{\eta}, \mathbb{P}y(n_0T; 0, h, u^h) \rangle = \langle \mathbb{P}^*\hat{\eta}, y(n_0T; 0, h, u^h) \rangle = 0 \quad \text{for all } \hat{\eta} \in H.$$

So $\mathbb{P}y(n_0T; 0, h, u^h) = 0$, which leads to (2.61).

From Step 1–Step 4, we end the proof of $(b) \Leftrightarrow (c)$.

We then show $(c) \Leftrightarrow (d)$. To show that $(c) \Rightarrow (d)$, we suppose that Z satisfies (c). Let μ and ξ satisfy the conditions in (d) with the aforementioned Z. Then by (1.44), we find that $\xi \in \tilde{H}_1^C$. Hence, we can write $\xi \triangleq \xi_1 + i\xi_2$ with $\xi_1, \xi_2 \in \tilde{H}_1$. By (1.43), we have $\xi_1, \xi_2 \in \mathbb{P}^* H_1$. By the second condition in (d), we see that

$$(B(t)|_Z)^{*C} \psi^\xi(t) = 0 \quad \text{for a.e. } t \in (0, T).$$

Then by (2.60) and the first condition in (d), we find that for all $t \in [0, T]$ and $k = 1, \dots, n_0$,

$$\psi_{n_0}^\xi((k-1)T + t) = \psi^{\mu^{n_0-k}\xi}(t) = \mu^{n_0-k}\psi^\xi(t).$$

Since $\psi_{n_0}^\xi(\cdot) = \psi_{n_0}^{\xi_1}(\cdot) + i\psi_{n_0}^{\xi_2}(\cdot)$, the above two equations yield that

$$(B(\cdot)|_Z)^* \psi_{n_0}^{\xi_1}(\cdot) + i\,(B(\cdot)|_Z)^* \psi_{n_0}^{\xi_2}(\cdot) = (B(\cdot)|_Z)^* \psi_{n_0}^\xi(\cdot) = 0 \quad \text{over} \quad (0, n_0T).$$

Since $\xi_1, \xi_2 \in \mathbb{P}^* H_1$, the above-mentioned equation, along with (c), leads to $\xi_1 = \xi_2 = 0$, i.e., $\xi = 0$. Hence, Z satisfies (d). Thus, we have proved that $(c) \Rightarrow (d)$.

To show that $(d) \Rightarrow (c)$, we suppose that $Z \subseteqq U$ satisfies (d). In order to show that Z satisfies (c), it suffices to prove that

$$\hat{\xi} \in (\mathbb{P}^* H_1)^C \text{ and } (B(\cdot)|_Z)^{*C} \psi_{n_0}^{\hat{\xi}}(\cdot) = 0 \text{ over } (0, n_0T) \Rightarrow \hat{\xi} = 0. \qquad (2.73)$$

Notice that $(\mathbb{P}^* H_1)^C = \tilde{H}_1^C$ and $\dim \tilde{H}_1^C = n_0$ (see Proposition 1.5). Simply write

$$\mathcal{Q} \triangleq \mathcal{P}^{*C}|_{\tilde{H}_1^C} \in \mathcal{L}(\tilde{H}_1^C) \quad \text{and} \quad B_1(\cdot) \triangleq ((B(\cdot)|_Z)^{*C})|_{(0,T)} \in L^2(0, T; \mathcal{L}(H, Z)).$$

By Proposition 1.5 and (1.37), we have that $\sigma(\mathcal{Q}) = \{\bar{\lambda}_j\}_{j=1}^{\hat{n}}$; l_j is the algebraic multiplicity of $\bar{\lambda}_j$. Hence, $p(\lambda) \triangleq \prod_{j=1}^{\hat{n}} (\lambda - \bar{\lambda}_j)^{l_j}$ is the characteristic polynomial of $\mathcal{Q}$. Write $\hat{l}_j$ for the geometric multiplicity of $\bar{\lambda}_j$. Clearly, $\hat{l}_j \le l_j$ for all j. Let

$\beta \triangleq \{\beta_1, \ldots, \beta_{n_0}\}$ be a basis of $(\mathbb{P}^* H)^C = \tilde{H}_1^C$ so that

$$\mathcal{Q}(\beta_1, \ldots, \beta_{n_0}) = J(\beta_1, \ldots, \beta_{n_0}). \tag{2.74}$$

Here J is the Jordan matrix: $\mathrm{diag}\big\{J_{11}, \ldots, J_{1\hat{l}_1}, J_{21}, \ \ldots, \ J_{2\hat{l}_2}, \ldots, J_{\hat{n}1}, \ \ldots, \ J_{\hat{n}\hat{l}_{\hat{n}}}\big\}$ with

$$J_{jk} = \begin{pmatrix} \bar{\lambda}_j & 1 & & \\ & \ddots & \ddots & \\ & & \ddots & 1 \\ & & & \bar{\lambda}_j \end{pmatrix} \quad \text{a } d_{jk} \times d_{jk} \text{ matrix,}$$

where $j = 1, \ldots, \hat{n}$, $k = 1, \ldots, \hat{l}_j$, and for each j, $\{d_{jk}\}_{k=1}^{\hat{l}_j}$ is decreasing. It is clear that $\sum_{k=1}^{\hat{l}_j} d_{jk} = l_j$ for each $j = 1, \ldots, \hat{n}$, and $\sum_{j=1}^{\hat{n}} \sum_{k=1}^{\hat{l}_j} d_{jk} = n_0$. We rewrite the basis β as

$$\beta \triangleq \big\{\xi_{111}, \ldots, \xi_{11d_{11}}, \xi_{1\hat{l}_1 1}, \ldots, \xi_{1\hat{l}_1 d_{1\hat{l}_1}}, \ldots, \xi_{\hat{n}11}, \ldots, \xi_{\hat{n}1d_{\hat{n}1}}, \xi_{\hat{n}\hat{l}_{\hat{n}}1}, \ldots, \xi_{\hat{n}\hat{l}_{\hat{n}} d_{\hat{n}\hat{l}_{\hat{n}}}}\big\}.$$

Then by (2.74), one can easily check that for each $j \in \{1, \ldots, \hat{n}\}$ and $k \in \{1, \ldots, \hat{l}_j\}$,

$$\big(\bar{\lambda}_j I - \mathcal{Q}\big)^q \xi_{jkr} = \begin{cases} \xi_{jk(r-q)} & \text{when } r > q, \\ 0 & \text{when } r \le q. \end{cases} \tag{2.75}$$

Now we assume $\hat{\xi}$ satisfies the conditions on the left side of (2.73). Since $\hat{\xi} \in (\mathbb{P}^* H_1)^C = \tilde{H}_1^C$, there is a vector

$$\big(C_{111}, \ldots, C_{11d_{11}}, C_{1\hat{l}_1 1}, \ldots, C_{1\hat{l}_1 d_{1\hat{l}_1}}, \ldots, C_{\hat{n}11}, \ldots, C_{\hat{n}1d_{\hat{n}1}}, C_{\hat{n}\hat{l}_{\hat{n}}1}, \ldots, C_{\hat{n}\hat{l}_{\hat{n}} d_{\hat{n}\hat{l}_{\hat{n}}}}\big)^* \in \mathbb{C}^{n_0},$$

so that

$$\hat{\xi} = \sum_{j=1}^{\hat{n}} \sum_{k=1}^{\hat{l}_j} \sum_{r=1}^{d_{jk}} C_{jkr} \xi_{jkr}. \tag{2.76}$$

From (2.59) and the second condition on the left side of (2.73), it follows that for each $m \in \{0, \ldots, n_0 - 1\}$, $B_1(\cdot)\psi_{n_0}^{\hat{\xi}}(\cdot)\big|_{((n_0-m-1)T, (n_0-m)T)} = 0$, that is,

$$\sum_{j=1}^{\hat{n}} \sum_{k=1}^{\hat{l}_j} \sum_{r=1}^{d_{jk}} C_{jkr} B_1(t) \psi^{\mathcal{Q}^m \xi_{jkr}}(t) = 0 \quad \text{for a.e. } t \in (0, T).$$

From this, we see that for any polynomial g with $\mathrm{degree}(g) \le n_0 - 1$,

$$\sum_{j=1}^{\hat{n}}\sum_{k=1}^{\hat{l}_j}\sum_{r=1}^{d_{jk}} C_{jkr}B_1(\cdot)\psi^{g(\mathcal{Q})\xi_{jkr}} = 0 \text{ over } (0,T). \tag{2.77}$$

Given $\widetilde{j} \in \{1,\ldots,\hat{n}\}$, let

$$p_{\widetilde{j}}(\lambda) = \prod_{j=1, j\neq\widetilde{j}}^{\hat{n}} (\lambda - \bar{\lambda}_j)^{l_j}.$$

By taking $g(\lambda) = \lambda^m p_{\widetilde{j}}(\lambda)$, with $m = 0, 1, \ldots, l_{\widetilde{j}} - 1$, in (2.77), we find that

$$\sum_{j=1}^{\hat{n}}\sum_{k=1}^{\hat{l}_j}\sum_{r=1}^{d_{jk}} C_{jkr}B_1(\cdot)\psi^{\mathcal{Q}^m p_{\widetilde{j}}(\mathcal{Q})\xi_{jkr}}(\cdot) = 0 \text{ over } (0,T), \text{ when } m \in \{0,1,\ldots,l_{\widetilde{j}}-1\}.$$

By (2.75), we see that $p_{\widetilde{j}}(\mathcal{Q})\xi_{jkr} = 0$, when $j \in \{1,\ldots,\hat{n}\}$, $j \neq \widetilde{j}$, $k \in \{1,\ldots,\hat{l}_j\}$ and $r \in \{1,\ldots,d_{jk}\}$. The above two equations imply that for each $m \in \{0,1,\ldots,l_{\widetilde{j}}-1\}$,

$$\sum_{k=1}^{\hat{l}_{\widetilde{j}}}\sum_{r=1}^{d_{\widetilde{j}k}} C_{\widetilde{j}kr}B_1(\cdot)\psi^{\mathcal{Q}^m p_{\widetilde{j}}(\mathcal{Q})\xi_{\widetilde{j}kr}}(\cdot) = 0 \text{ over } (0,T),$$

from which, it follows that for any polynomial f with degree$(f) \leq l_{\widetilde{j}} - 1$,

$$\sum_{k=1}^{\hat{l}_{\widetilde{j}}}\sum_{r=1}^{d_{\widetilde{j}k}} C_{\widetilde{j}kr}B_1(\cdot)\psi^{f(\mathcal{Q})p_{\widetilde{j}}(\mathcal{Q})\xi_{\widetilde{j}kr}}(\cdot) = 0 \text{ over } (0,T). \tag{2.78}$$

Given $m \in \{0,1,2,\cdots,l_{\widetilde{j}}-1\}$, since $p_{\widetilde{j}}(\lambda)$ and $(\lambda - \bar{\lambda}_{\widetilde{j}})^{m+1}$ are coprime, there are polynomials $g_m^1(\lambda)$ and $g_m^2(\lambda)$ with degree$(g_m^1) \leq m$ and degree$(g_m^2) \leq$ degree $(p_{\widetilde{j}}) - 1$, respectively, so that

$$g_m^1(\lambda)p_{\widetilde{j}}(\lambda) + g_m^2(\lambda)(\lambda - \bar{\lambda}_{\widetilde{j}})^{m+1} \equiv 1.$$

Thus, for all $m \in \{0,1,\cdots,l_{\widetilde{j}}-1\}$, $k \in \{1,2,\cdots,\hat{l}_{\widetilde{j}}\}$, and $r \in \{1,2,\cdots,d_{\widetilde{j}k}\}$,

$$(\mathcal{Q}-\bar{\lambda}_{\widetilde{j}}I)^{l_{\widetilde{j}}-m-1}g_m^1(\mathcal{Q})\mathbb{P}_{\widetilde{j}}(\mathcal{Q})\xi_{\widetilde{j}kr}+g_m^2(\mathcal{Q})(\mathcal{Q}-\bar{\lambda}_{\widetilde{j}}I)^{l_{\widetilde{j}}}\xi_{\widetilde{j}kr} \equiv (\mathcal{Q}-\bar{\lambda}_{\widetilde{j}}I)^{l_{\widetilde{j}}-m-1}\xi_{\widetilde{j}kr}. \tag{2.79}$$

By (2.75), we have that

$$(\hat{\mathcal{Q}} - \bar{\lambda}_{\widetilde{j}}I)^{l_{\widetilde{j}}}\xi_{\widetilde{j}kr} = 0 \text{ for all } k \in \{1,2,\cdots,\hat{l}_{\widetilde{j}}\},\ r \in \{1,2,\cdots,d_{\widetilde{j}k}\}. \tag{2.80}$$

Taking $f(\lambda) = (\lambda - \bar{\lambda}_{\tilde{j}})^{l_{\tilde{j}}-m-1} g_m^1(\lambda)$, with $m = 0, \ldots, l_{\tilde{j}} - 1$, in (2.78), using (2.79) and (2.80), we find that for each $m \in \{0, 1, \ldots, l_{\tilde{j}} - 1\}$,

$$\sum_{k=1}^{\hat{l}_{\tilde{j}}} \sum_{r=1}^{d_{\tilde{j}k}} C_{\tilde{j}kr} B_1(\cdot) \psi^{(\mathcal{Q}-\bar{\lambda}_{\tilde{j}} I)^m \xi_{\tilde{j}kr}}(\cdot) = 0 \quad \text{over} \quad (0, T). \tag{2.81}$$

Now we are on the position to show that

$$C_{\tilde{j}kr} = 0 \quad \text{for all} \quad k \in \{1, 2, \cdots, \hat{l}_{\tilde{j}}\},\ r \in \{1, \ldots, d_{\tilde{j}k}\}, \tag{2.82}$$

which leads to $\hat{\xi} = 0$ because of (2.76). For this purpose, we write

$$K_{\tilde{j}}^m = \left\{ k \in \{1, 2, \ldots, \hat{l}_{\tilde{j}}\} \mid d_{\tilde{j}k} > m \right\}, \quad m = 0, 1, \ldots, l_{\tilde{j}} - 1.$$

One can easily check that (2.82) is equivalent to

$$\mathscr{C}_{\hat{m}} \triangleq \left\{ C_{\tilde{j}k\hat{m}},\ \ k \in K_{\tilde{j}}^{\hat{m}-1} \right\} = \{0\} \quad \text{for all} \quad \hat{m} \in \{1, \ldots, d_{\tilde{j}1}\}. \tag{2.83}$$

We will use the mathematical induction with respect to $\hat{m}$ to prove (2.83). (Notice that $d_{\tilde{j}k}$ is decreasing with respect to k.) First of all, we let

$$\mathbb{Q}_{\tilde{j}}^m(\lambda) = \left(\bar{\lambda}_{\tilde{j}} - \lambda\right)^m, \quad m = 0, 1, \ldots, l_{\tilde{j}} - 1, \tag{2.84}$$

In the case that $\hat{m} = d_{\tilde{j}1}$, it follows from (2.84) and (2.75) that

$$\mathbb{Q}_{\tilde{j}}^{\hat{m}-1}(\mathcal{Q}) \xi_{\tilde{j}k\hat{m}} = \left(\bar{\lambda}_{\tilde{j}} I - \mathcal{Q}\right)^{\hat{m}-1} \xi_{\tilde{j}k\hat{m}} = \xi_{\tilde{j}k1}, \quad \text{when } k \in K_{\tilde{j}}^{\hat{m}-1},$$

and

$$\mathbb{Q}_{\tilde{j}}^{\hat{m}-1}(\mathcal{Q}) \xi_{\tilde{j}kr} = 0, \quad \text{when } k \in K_{\tilde{j}}^{\hat{m}-1},\ r < \hat{m};\ \text{or } k \notin K_{\tilde{j}}^{\hat{m}-1},\ r \in \{1, \ldots, d_{\tilde{j}k}\}.$$

These, alone with (2.81) (where $m = \hat{m} - 1$), imply that

$$\sum_{k \in K_{\tilde{j}}^{\hat{m}-1}} C_{\tilde{j}k\hat{m}} B_1(\cdot) \psi^{\xi_{\tilde{j}k1}}(\cdot) = 0 \quad \text{over} \quad (0, T).$$

Let

$$\bar{\xi}_{\hat{m}} \triangleq \sum_{k \in K_{\tilde{j}}^{\hat{m}-1}} C_{\tilde{j}k\hat{m}} \xi_{\tilde{j}k1} \quad \text{with} \quad \hat{m} = 1, \ldots, d_{\tilde{j}1}.$$

Then, it holds that

$$B_1(\cdot)\psi^{\bar{\xi}_{\hat{m}}}(\cdot) = 0 \ \text{ over } \ (0, T). \tag{2.85}$$

Since for each $k \in \{1, \ldots, \hat{l}_{\tilde{j}}\}$, $\xi_{\tilde{j}k1}$ is an eigenfunction of $\mathscr{Q}$ with respect to the eigenvalue $\bar{\lambda}_{\tilde{j}}$, it follows from the definition of $\bar{\xi}_{\hat{m}}$ that

$$(\bar{\lambda}_{\tilde{j}} I - \mathscr{Q})\bar{\xi}_{\hat{m}} = 0.$$

This, along with (2.85) and (d), yields that $\bar{\xi}_{\hat{m}} = 0$, i.e., $\bar{\xi}_{d_{\tilde{j},1}} = 0$, which leads to $\mathscr{C}_{d_{\tilde{j}1}} = 0$ because of the linear independence of the group $\{\xi_{\tilde{j}k1},\ k \in K_{\tilde{j}}^{\hat{m}-1}\}$. Hence, (2.83) holds when $\hat{m} = d_{\tilde{j}1}$. Suppose inductively that (2.83) holds when $\tilde{m} + 1 \le \hat{m} \le d_{\tilde{j}1}$ for some $\tilde{m} \in \{1, \ldots, d_{\tilde{j}1} - 1\}$, i.e.,

$$\mathscr{C}_{\hat{m}} = \{0\}, \quad \text{when } \ \tilde{m} + 1 \le \hat{m} \le d_{\tilde{j}1}. \tag{2.86}$$

We will show that (2.83) holds when $\hat{m} = \tilde{m}$. In fact, it follows from (2.75) that

$$\mathbb{Q}_{\tilde{j}}^{\tilde{m}-1}(\mathscr{Q})\xi_{\tilde{j}kr} = \begin{cases} \xi_{\tilde{j}k(r-\tilde{m}+1)}, & \text{when } k \in K_{\tilde{j}}^{\tilde{m}-1},\ r \ge \tilde{m}, \\ 0, & \text{when } k \in K_{\tilde{j}}^{\tilde{m}-1},\ r < \tilde{m}, \\ 0, & \text{when } k \notin K_{\tilde{j}}^{\tilde{m}-1},\ r \in \{1, \ldots, d_{\tilde{j}k}\}. \end{cases}$$

This, alone with (2.81) (where $m = \tilde{m} - 1$), indicates that

$$\sum_{k=1}^{\hat{l}_{\tilde{j}}} \sum_{r=1}^{d_{\tilde{j}k}} C_{\tilde{j}kr} B_1(\cdot)\psi^{p_{\tilde{j}}^{\tilde{m}-1}(\mathscr{Q})\xi_{\tilde{j}kr}}(\cdot) = \sum_{k \in K_{\tilde{j}}^{\tilde{m}-1}} \sum_{r=\tilde{m}}^{d_{\tilde{j}k}} C_{\tilde{j}kr} B_1(\cdot)\psi^{\xi_{\tilde{j}k(r-\tilde{m}+1)}}(\cdot) = 0 \ \text{ over } \ (0, T).$$

Then, by (2.86), we have that

$$\sum_{k \in K_{\tilde{j}}^{\tilde{m}-1}} C_{\tilde{j}k\tilde{m}} B_1(\cdot)\psi^{\xi_{\tilde{j}k1}}(\cdot) = 0 \ \text{ over } \ (0, T). \tag{2.87}$$

Let

$$\bar{\xi}_{\tilde{m}} \triangleq \sum_{k \in K_{\tilde{j}}^{\tilde{m}-1}} C_{\tilde{j}k\tilde{m}} \xi_{\tilde{j}k1}.$$

Then, it follows from (2.87) that

$$B_1(\cdot)\psi^{\bar{\xi}_{\tilde{m}}}(\cdot) = 0 \ \text{ over } \ (0, T). \tag{2.88}$$

Since for each $k \in \{1, \ldots, \hat{l}_{\tilde{j}}\}$, $\xi_{\tilde{j},k,1}$ is an eigenfunction of $\mathscr{Q}$ with respect to the eigenvalue $\bar{\lambda}_{\tilde{j}}$, it holds that $(\bar{\lambda}_{\tilde{j}} I - \mathscr{Q})\bar{\xi}_{\tilde{m}} = 0$. This, along with (2.88) and (d), yields that $\bar{\xi}_{\tilde{m}} = 0$. Hence, $\mathscr{C}_{\tilde{m}} = \{0\}$ because of the linear independence of the group

$\{\xi_{\tilde{j}k1},\ k \in K_{\tilde{j}}^{\tilde{m}-1}\}$. In summary, we conclude that $(d) \Rightarrow (c)$. This completes the proof of Theorem 2.1. □

2.3 Applications

Some applications of Theorem 2.1 will be given in this section.

2.3.1 *Feedback Realization in Finite Dimensional Subspaces*

When Eq. (1.1) is LPFS, can we find a finite dimensional subspace Z of U so that $Z \in \mathscr{U}^{FS}$? The answer is positive. This might help us to design a feedback law numerically. To prove the above-mentioned positive answer, the following lemma is needed.

Lemma 2.3 *For each subspace $Z \subseteqq U$, there is a finite dimensional subspace $\hat{Z} \subseteqq Z$ so that*

$$\hat{\mathscr{A}}_{n_0}^{Z} = \hat{\mathscr{A}}_{n_0}^{\hat{Z}} \quad and \quad \dim \hat{Z} \le n_0 \tag{2.89}$$

where $\hat{\mathscr{A}}_{n_0}^{Z}$ and $\hat{\mathscr{A}}_{n_0}^{\hat{Z}}$ are defined by (2.4), and n_0 is given by (1.17).

Proof We carry out the proof by two steps.

Step 1. To show that there is a finite-dimensional subspace $\tilde{Z}$ of U so that $\hat{\mathscr{A}}_{n_0}^{Z} = \hat{\mathscr{A}}_{n_0}^{\hat{Z}}$
Let Z be a subspace of U. Since $\hat{\mathscr{A}}_{n_0}^{Z}$ is a subspace of H_1 and $\dim H_1 = n_0 < \infty$ (see (1.22)), we can assume that $\dim \hat{\mathscr{A}}_{n_0}^{Z} \triangleq m \le n_0$. Write $\{\xi_1, \ldots, \xi_m\}$ for an orthonormal basis of $\hat{\mathscr{A}}_{n_0}^{Z}$. By (2.4) and (2.2), there are $u_j(\cdot) \in L^2(\mathbb{R}^+; Z)$, $j = 1, \ldots, m$, so that

$$\mathscr{L}_1 u_j = \xi_j \quad \text{for all} \quad j = 1, \ldots, m, \tag{2.90}$$

where $\mathscr{L}_1 : L^2(\mathbb{R}^+; Z) \to H_1$ is defined by

$$\mathscr{L}_1 u \triangleq \int_0^{n_0 T} \mathbb{P}\Phi(n_0 T, s) D(s) u(s) \mathrm{d}s, \ u \in L^2(\mathbb{R}^+; Z).$$

From the orthonormality, it follows that

$$\det\left(\langle \mathscr{L}_1 u_i, \mathscr{L}_1 u_j \rangle\right)_{ij} = 1 \neq 0. \tag{2.91}$$

By the definition of the Bochner integration (*see* [14]), for each $j \in \{1, \dots, m\}$, there is a sequence of simple functions, denoted by $\{v_j^k\}_{k=1}^{\infty}$, so that

$$\lim_{k\to\infty} \int_0^{n_0 T} \|v_j^k(s) - u_j(s)\|_U \mathrm{d}s = 0.$$

This, along with (2.91), yields that there is a k_0 such that $\det\Big(\langle \mathscr{L}_1 v_i^{k_0},\ \mathscr{L}_1 v_j^{k_0}\rangle\Big)_{ij} \neq 0$. Let

$$\eta_j = \mathscr{L}_1 v_j^{k_0}, \quad j = 1, \dots, m. \tag{2.92}$$

Then, $\{\eta_1, \dots, \eta_m\}$ is a linearly independent group in the subspace $\hat{\mathscr{A}}_{n_0}^{Z}$. Hence, $\{\eta_1, \dots, \eta_m\}$ is a basis of $\hat{\mathscr{A}}_{n_0}^{Z}$. Write

$$v_j^{k_0}(\cdot) = \sum_{l=1}^{k_j} \chi_{E_{jl}}(\cdot) z_{jl} \ \text{ over } \ (0, n_0 T), \quad j = 1, \dots, m, \tag{2.93}$$

with $z_{jl} \in Z$, E_{jl} measurable sets in $(0, n_0 T)$ and $\chi_{E_{jl}}$ the characteristic function of E_{jl}. Let

$$\tilde{Z} = \text{span}\left\{z_{11}, \dots, z_{1k_1}, z_{21}, \dots, z_{2k_2}, \dots, z_{m1}, \dots, z_{mk_m}\right\}.$$

Clearly, $\tilde{Z}$ is a finite-dimensional subspace of Z and all $v_j^{k_0}(\cdot)$, $j = 1, \dots, m$, (given by (2.93)) belong to $L^2(\mathbb{R}^+; \tilde{Z})$. The later, along with (2.92), yields that $\eta_j \in \hat{V}_{n_0}^{\tilde{Z}}$ for each $j = 1, \dots, m$. Hence,

$$\hat{\mathscr{A}}_{n_0}^{Z} \supseteq \hat{\mathscr{A}}_{n_0}^{\tilde{Z}} \supseteq \text{span}\{\eta_1, \dots, \eta_m\} = \hat{\mathscr{A}}_{n_0}^{Z}.$$

This leads to $\hat{\mathscr{A}}_{n_0}^{Z} = \hat{\mathscr{A}}_{n_0}^{\tilde{Z}}$.

Step 2. To show that there is a subspace $\hat{Z}$ of $\tilde{Z}$ *(which is constructed in Step 1) such that* $\hat{\mathscr{A}}_{n_0}^{\hat{Z}} = \hat{\mathscr{A}}_{n_0}^{\tilde{Z}}$ and $\dim \hat{Z} \leq n_0$

Write $\{\zeta_1, \zeta_2, \dots, \zeta_{l_0}\}$ for an orthonormal basis of $\tilde{Z}$. Then it holds that

$$\hat{\mathscr{A}}_{n_0}^{\tilde{Z}} = \sum_{j=1}^{l_0} \hat{\mathscr{A}}_{n_0}^{Z_j}, \quad \text{with} \quad Z_j = \text{span}\{\zeta_j\}, \quad j = 1, 2, \dots, l_0. \tag{2.94}$$

For each $j \in \{1, \dots, l_0\}$, let $\{\eta_{j1}, \eta_{j2}, \dots, \eta_{jk_j}\}$ be a basis of $\hat{\mathscr{A}}_{n_0}^{Z_j}$. Denote by S a maximal independent group of the set:

$$\{\eta_{11}, \dots, \eta_{1,k_1}, \eta_{21}, \dots, \eta_{2,k_2}, \dots, \eta_{l_0,1}, \dots, \eta_{l_0,k_{l_0}}\}.$$

Then

$$\text{span}\left\{\eta \mid \eta \in S\right\} = \hat{\mathscr{A}}^{\tilde{Z}}_{n_0}, \tag{2.95}$$

and S contains $m \leq n_0$ elements. Let

$$J = \left\{j = 1, \ldots, l_0 \mid \eta_{jk} \in S \text{ for some } k \in \{1, \ldots, k_j\}\right\}.$$

Then the number of elements contained in J equals to or is less than $m \leq n_0$. Set $\hat{Z} = \sum_{j\in J} Z_j$. Then it holds that $\hat{\mathscr{A}}^{\hat{Z}}_{n_0} \subseteq \hat{\mathscr{A}}^{\tilde{Z}}_{n_0}$ and span $\left\{\eta \mid \eta \in S\right\} \subseteq \sum_{j\in J} \hat{\mathscr{A}}^{Z_j}_{n_0} = \hat{\mathscr{A}}^{\hat{Z}}_{n_0}$. These, together with (2.95), yield that $\hat{\mathscr{A}}^{\hat{Z}}_{n_0} = \hat{\mathscr{A}}^{\tilde{Z}}_{n_0}$. Because of the number of elements contained in J equals to or is less than $m \leq n_0$, it holds that dim$\hat{Z} \leq m \leq n_0$. This completes the proof. □

Now a main result of this subsection is presented by the following theorem.

Theorem 2.2 *Equation (1.1) is LPFS if and only if it is LPFS with respect to a subspace Z of U with dim$Z \leq n_0$.*

Proof Clearly, it suffices to show the *only if* part. Assume that Eq. (1.1) is LPFS. By the equivalence of (a) and (b) in Theorem 2.1, we find that $\hat{\mathscr{A}}^{U}_{n_0} = H_1$. Meanwhile, by Lemma 2.3, there is a subspace of $\hat{Z}$ of U so that $\hat{\mathscr{A}}^{U}_{n_0} = \hat{\mathscr{A}}^{\hat{Z}}_{n_0}$ with dim$\hat{Z} \leq n_0$. Thus, we find that $\hat{\mathscr{A}}^{\hat{Z}}_{n_0} = H_1$. This, along with the equivalence of (a) and (b) in Theorem 2.1, indicates that Eq. (1.1) is LPFS with respect to $\hat{Z}$. This ends the proof. □

2.3.2 Applications to Heat Equations

In this subsection, we will present some applications of Theorem 2.1 to heat equations with time-periodic potentials.

Let Ω be a bounded domain in $\mathbb{R}^d$ $(d \geq 1)$ with a C^2-smooth boundary $\partial\Omega$. Write $Q \triangleq \Omega \times \mathbb{R}^+$ and $\sum \triangleq \partial\Omega \times \mathbb{R}^+$. Let $\omega \subseteq \Omega$ be a non-empty open subset with its characteristic function χ_ω. Let $T > 0$ and $a \in C(\bar{Q})$ be T-periodic (with respect to the time variable t), i.e., for each $t \in \mathbb{R}^+$, $a(\cdot, t) = a(\cdot, t + T)$ over Ω. One can easily check that the function a can be treated as a T-periodic function in $L^1_{loc}(\mathbb{R}^+; \mathscr{L}(L^2(\Omega)))$. Consider the following controlled heat equation:

$$\begin{cases} \partial_t y(x,t) - \Delta y(x,t) + a(x,t)y(x,t) = \chi_\omega(x)u(x,t) & \text{in } Q, \\ y(x,t) = 0 & \text{on } \sum, \end{cases} \tag{2.96}$$

where $u \in L^2(\mathbb{R}^+; L^2(\Omega))$. Given $y_0 \in L^2(\Omega)$ and $u \in L^2(\mathbb{R}^+; L^2(\Omega))$, Eq. (2.96), with the initial condition that $y(x, 0) = y_0(x)$, has a unique solution $y(\cdot; 0, y_0, u)$ in the space $C(\mathbb{R}^+; L^2(\Omega))$. Let $H = U = L^2(\Omega)$ and $A = \Delta$ with $\mathscr{D}(A) = H^1_0(\Omega) \bigcap$

$H_2(\Omega)$. Define, for each $t \in \mathbb{R}^+$, $D(t)$: $H \to H$ by $D(t)z(x) = -a(x,t)z(x)$, $x \in \Omega$, and $B(t)$: $U \to H$ by $B(t)v(x) = \chi_\omega(x)v(x)$, $x \in \Omega$. Clearly, A generates a compact semigroup on $L^2(\Omega)$ and both $D(\cdot) \in L^1_{loc}(\mathbb{R}^+; \mathscr{L}(L^2(\Omega)))$ and $B(\cdot) \in L^\infty(\mathbb{R}^+; \mathscr{L}(U; H))$ are T-periodic. Thus, we can study Eq. (2.96) under the framework (1.1). Write Ψ_a for the evolution generated by $A + D(\cdot)$. We use notations n_0, $\mathbb{P}$, H_j (with $j = 1, 2$), $\mathscr{A}_k^Z$ and $\hat{\mathscr{A}}_k^Z$ (with $k \in \mathbb{N}$) to denote the same subjects as those introduced in the first section of this chapter.

Corollary 2.1 *Equation (2.96) is LPFS with respect to a subspace Z of $L^2(\Omega)$ with $dim Z \leq n_0$.*

Proof We will provide two ways to show that Eq. (2.96) is LPFS. We first use the equivalence $(a) \Leftrightarrow (c)$ in Theorem 2.1. In fact, $\psi(\cdot) \triangleq \Psi_a(n_0T, \cdot)^*\xi$ with $\xi \in H$ is the solution to the equation:

$$\begin{cases} \partial_t\psi(x,t) + \Delta\psi(x,t) - a(x,t)\psi(x,t) = 0 & \text{in } \Omega \times (0, n_0T), \\ \psi(x,t) = 0 & \text{on } \partial\Omega \times (0, n_0T), \\ \psi(x, n_0T) = \xi(x) & \text{in } \Omega. \end{cases} \tag{2.97}$$

Moreover, we have that

$$B(t)\eta = \chi_\omega\eta \ \text{ for any } \ \eta \in H \ \text{ and } \ t \in [0, T]. \tag{2.98}$$

These, along with the unique continuation property of parabolic equations established in [60] (see also [74, 75]), lead to the condition (c) in Theorem 2.1 for the current case. Then, according to the equivalence $(a) \Leftrightarrow (c)$ in Theorem 2.1, Eq. (2.96) is LPFS.

We next use the equivalence $(a) \Leftrightarrow (b)$ in Theorem 2.1. Without loss of generality, we can assume that $n_0 \geq 1$, for otherwise Eq. (2.96), with the null control, is stable. When $n_0 \geq 1$, we have $H_1 \neq \{0\}$ and $||P|| > 0$. Write $\{\xi_1, \ldots, \xi_{n_0}\}$ for an orthonormal basis of H_1. By the approximate controllability of the heat equation (*see* [33]), $\mathscr{A}_1^U$ is dense in H. Thus there are η_j, $j = 1 \ldots, n_0$, in $\mathscr{A}_1^U$ so that

$$\|\eta_j - \xi_j\| \leq \frac{1}{16n_0\|\mathbb{P}\|} \ \text{ for all } \ j = 1, \ldots, n_0. \tag{2.99}$$

Since P is a projection from H onto H_1, we have that $\mathbb{P}\xi_j = \xi_j$ for all $j = 1, \ldots, n_0$. This, along with (2.99), yields that for each $j \in \{1, \ldots, n_0\}$,

$$\|\mathbb{P}\eta_j\| \leq \|\xi_j\| + \|\mathbb{P}\|\|\eta_j - \xi_j\| \leq 1 + \frac{1}{16n_0}; \tag{2.100}$$

and

$$\langle\mathbb{P}\eta_j, \xi_j\rangle \geq 1 - \frac{1}{16n_0}. \tag{2.101}$$

Since $\mathbb{P}\eta_j \in H_1$ and $\{\xi_k\}_{k=1}^{n_0}$ is an orthonormal basis of H_1, we find that

$$\|\mathbb{P}\eta_j\|^2 = \sum_{k=1}^{n_0} |\langle \mathbb{P}\eta_j, \xi_k\rangle|^2, \quad \text{when} \quad j = 1, \ldots, n_0. \tag{2.102}$$

From (2.102), (2.100) and (2.101), it follows that for each $j \in \{1, \ldots, n_0\}$,

$$\begin{aligned}\sum_{k\neq j} \big|\langle \mathbb{P}\eta_j, \xi_k\rangle\big| &\leq (n_0 - 1)^{1/2}\Big(\sum_{k\neq j} \big|\langle \mathbb{P}\eta_j, \xi_k\rangle\big|^2\Big)^{1/2} \\ &= (n_0 - 1)^{1/2}\big(\|\mathbb{P}\eta_j\|^2 - |\langle \mathbb{P}\eta_j, \xi_j\rangle|^2\big)^{1/2} \\ &\leq n_0^{1/2}\big((1 + 1/(16n_0))^2 - (1 - 1/(16n_0))^2\big)^{1/2} = 1/2.\end{aligned}$$

This, together with (2.101), indicates that

$$\langle \mathbb{P}\eta_j, \xi_j\rangle \geq 1 - 1/(16n_0) > 1/2 \geq \sum_{k\neq j} \big|\langle \mathbb{P}\eta_j, \xi_k\rangle\big|, \; j = 1, 2, \cdots, n_0. \tag{2.103}$$

We claim that $\{\mathbb{P}\eta_1, \ldots, \mathbb{P}\eta_{n_0}\}$ is a linearly independent group. In fact, suppose that

$$\sum_{j=1}^{n} c_j \mathbb{P}\eta_j = 0 \;\text{ for some }\; c_1, \ldots, c_{n_0} \in \mathbb{R}. \tag{2.104}$$

Write

$$\hat{A} \triangleq (\langle \mathbb{P}\eta_j, \xi_k\rangle)_{j,k} \in \mathbb{R}^{n_0 \times n_0} \quad \text{and} \quad \hat{c} \triangleq (c_1, \ldots, c_{n_0})^* \in \mathbb{R}^{n_0}.$$

By (2.103), the matrix $\hat{A}$ is diagonally dominant, hence it is invertible. Then, from (2.104), it follows that $\hat{A}^*\hat{c} = 0$, which implies $\hat{c} = 0$. Hence, $\mathbb{P}\eta_1, \ldots, \mathbb{P}\eta_{n_0}$ are linearly independent.

Since $\dim H_1 = n_0$, it follows that

$$span\{\mathbb{P}\eta_1, \ldots, \mathbb{P}\eta_{n_0}\} = H_1.$$

Therefore, we have that

$$H_1 \supseteq \hat{V}_{n_0}^U \supseteq \hat{V}_1^U = \mathbb{P}V_1^U \supseteq \text{span}\{\mathbb{P}\eta_1, \cdots, \mathbb{P}\eta_{n_0}\} = H_1,$$

from which, it follows that $H_1 = \hat{V}_{n_0}^U$. This, along with the equivalence of (a) and (b) in Theorem 2.1, indicates that Eq. (2.96) is LPFS.

Finally, according to Theorem 2.2, there is a subspace Z of $L^2(\Omega)$ with $\dim Z \leq n_0$ so that Eq. (2.96) is LPFS with respect to Z. This completes the proof. □

Corollary 2.2 *Equation (2.96) is LPFS with respect to the subspace* $\mathbb{P}^* H$.

Proof Let $Z = \mathbb{P}^* H$. By the equivalence between (a) and (d) in Theorem 2.1, it suffices to show that Z satisfies (d), i.e.,

$$\left.\begin{array}{l} \mu \notin \mathbb{B},\ \xi \in H^C,\ \left(\mu I - \mathscr{P}^{*C}\right)\xi = 0, \\ \left(B(\cdot)\big|_Z\right)^{*C} \Psi_a(T,\cdot)^{*C}\xi = 0 \ \text{ over } \ (0,T) \end{array}\right\} \Rightarrow \xi = 0. \tag{2.105}$$

Suppose that μ and ξ satisfy the conditions on the left side of (2.105). Write $\xi = \xi_1 + i\xi_2$ where $\xi_1, \xi_2 \in H$. Then, we have that

$$\left(B(\cdot)\big|_Z\right)^* \Psi_a(T,\cdot)^*\xi_j = 0 \ \text{ over } \ (0,T); \quad j = 1,2.$$

Since $\psi_j(\cdot) \triangleq \Psi_a(T,\cdot)^*\xi_j$ (with $j = 1, 2$) is the solution to the Eq. (2.97) (where $n_0 T$ and ξ are replaced by T and ξ_j respectively), $\Psi_a(T,\cdot)^*\xi_j$ is continuous on $[0, T]$ and $B(t)$ is independent of t, the above yields that

$$\left(B(0)\big|_Z\right)^* \Psi_a(T,0)^*\xi_j = 0, \quad \text{with } \ j = 1,2.$$

Since $\mathbb{P}^*\eta \in \mathbb{P}^* H$ for each $\eta \in H$, the above yields that

$$\begin{aligned} &\left\langle \left(B(0)\big|_Z\right)^* \Psi_a(T,0)^*\xi_j,\ \mathbb{P}^*\eta \right\rangle = \left\langle \Psi_a(T,0)^*\xi_j,\ \left(B(0)\big|_Z\right)\mathbb{P}^*\eta \right\rangle \\ &= \left\langle \Psi_a(T,0)^*\xi_j,\ \chi_\omega \mathbb{P}^*\eta \right\rangle = \left\langle \mathbb{P}\chi_\omega \Psi_a(T,0)^*\xi_j,\ \eta \right\rangle,\ j = 1,2. \end{aligned}$$

Hence, we have that

$$\mathbb{P}\chi_\omega \Psi_a(T,0)^*\xi_j = 0,\ j = 1,2,$$

from which, it follows that

$$\langle \mathbb{P}^*\Psi_a(T,0)^*\xi_j,\ \chi_\omega\Psi_a(T,0)^*\xi_j\rangle = \langle \Psi_a(T,0)^*\xi_j,\ \mathbb{P}\chi_\omega\Psi_a(T,0)^*\xi_j\rangle = 0,\ j = 1,2. \tag{2.106}$$

Two facts are as follows. First, it follows from (1.23) that

$$\mathbb{P}^*\Psi_a(T,0)^*\xi_j = \Psi_a(T,0)^*\mathbb{P}^*\xi_j,\ j = 1,2. \tag{2.107}$$

Second, by (1.44), (1.43), and the first three conditions on the left side of (2.105), we have $\xi \in \tilde{H}_1^C$. Since $\mathbb{P}^* = \tilde{\mathbb{P}}$ and $\tilde{\mathbb{P}}$ is a projection from H to $\tilde{H}_1$ (see Proposition 1.5), we see that $\mathbb{P}^* : H \to \tilde{H}_1$ is a projection. Hence, $\mathbb{P}^{*C} : H^C \to \tilde{H}_1^C$ is a projection. These two facts yields that $\mathbb{P}^{*C}\xi = \xi$, from which, it follows that $\mathbb{P}^*\xi_j = \xi_j$, $j = 1,2$. This along with (2.106) and (2.107), indicates that $\left\| \chi_\omega \Psi_a(T,0)^{*C}\xi \right\| = 0$. By the unique continuation property of parabolic equations established in [60] (see also [74, 75]), we find that $\xi_j = 0$, $j = 1,2$, which leads to $\xi = 0$. This completes the proof. □

We next introduce a controlled heat equation which is not LPFS. Write λ_1 and λ_2 for the first and the second eigenvalues of the operator $-\Delta$ with $\mathscr{D}(-\Delta) = H_0^1(\Omega) \cap H^2(\Omega)$, respectively. Let ξ_j, $j = 1, 2$, be an eigenfunction corresponding to λ_j. Consider the following heat equation:

$$\begin{cases} \partial_t y(x,t) - \Delta y(x,t) - \lambda_2 y(x,t) = \langle u(t), \xi_1 \rangle \xi_1(x) & \text{in } Q, \\ y(x,t) = 0 & \text{on } \sum. \end{cases} \tag{2.108}$$

where $u(\cdot) \in L^2(\mathbb{R}^+; L^2(\Omega))$. By a direct calculation, one has that $V_{n_0} = \text{span}\{\xi_1\}$ and $H_1 \supseteq \text{span}\{\xi_1, \xi_2\}$. These, along with $(a) \Leftrightarrow (b)$ in Theorem 2.1, indicates that (2.108) is not LPFS.

We end this subsection with the following note: It should be an interesting problem how to find a finitely dimensional subspace Z from $\mathscr{U}^{FS}$ so that it has the minimal dimension. (Here, $\mathscr{U}^{FS}$ is given by (2.1)) In general, we are not able to solve this problem. However, in some cases, it can be done. In Example 4.2, by applying Theorem 2.1, as well as Theorem 2.2 and Corollary 2.1, to a controlled heat equation, we solved this problem. From this point of view, Example 4.2 is also an application of Theorem 2.1 to controlled heat equations. The reasons that we put this example at the end of the last section of Chap. 4 are as follows: First, this problem can be understood as designs of a kind of simple control machines in infinitely dimensional cases. Second, the problems of designs of some simple control machines in finitely dimensional settings will be introduced in Chap. 4.

Miscellaneous Notes

There have been studies on equivalence conditions of periodic feedback stabilization for linear periodic evolution systems. In [31, 67], the authors established an equivalent condition on stabilizability for linear time-periodic parabolic equations with open-loop controls. Their equivalence (see Theorem 3.1 in [67] and Proposition 3.1 in [31]) can be stated, under the framework of Sect. 1.1, as follows: the condition (d) (in Theorem 2.1 where $Z = U$) is equivalent to the statement that for any $h \in H$, there is a control $u^h(\cdot) \in C(\mathbb{R}^+; U)$, with $\sup_{t \in \mathbb{R}^+} \|e^{\bar{\delta} t} u^h(t)\|_U$ bounded (where $\bar{\delta}$ is given by (1.16)), so that the solution $y(\cdot; 0, h, u^h)$ is stable. Meanwhile, it was pointed out in [67] that when open-looped stabilization controls exist, one can construct a periodic feedback stabilization law through using a method provided in [30]. From this point of view, the equivalence $(a) \Leftrightarrow (d)$ in Theorem 2.1 has been built up in [31, 67], through a different way. The method to construct the stabilization feedback law in this chapter is different from that in [30]. Besides, we would like to mention the paper [7] where the authors built up a feedback law for some nonlinear time-periodic evolution systems.

Proposition 2.3 (see also Remark 2.1) is a byproduct of the main study in this chapter. It shows that when both $D(\cdot)$ and $B(\cdot)$ are time invariant, linear time-period functions $K(\cdot)$ will not aid the linear stabilization of Eq. (1.1), i.e., Eq. (1.1) is linear $\hat{T}$-periodic feedback stabilizable for some $\hat{T} > 0$ if and only if Eq. (1.1) is linear time invariant feedback stabilizable. On the other hand, when Eq. (1.1) is periodic time varying, linear time-periodic $K(\cdot)$ do aid in the linear stabilization of this equation.

The material of this chapter is adapted from [93].

Chapter 3
Criteria on Periodic Stabilization in Finite Dimensional Cases

Abstract This chapter presents two periodic stabilization criteria for linear periodic ODEs. One is an algebraic criterion which is related to the periodic map, while another is a geometric criterion which is connected with the null-controllable subspace of the system. Two kinds of periodic feedback laws are constructed. We approach the geometric criterion by a different way from that used in infinitely dimensional cases of Chap. 2.

Keywords Periodic Equations · Stabilization · Algebraic Criterion · Geometric Criterion · ODE

We will put our object in the framework of Chap. 1, with $\dim H = n$ and $\dim U = m$. Without loss of generality, we assume that $H = \mathbb{R}^n$ and $U = \mathbb{R}^m$. The controlled system (1.1) and (1.2) read respectively:

$$y'(s) = A(s)y(s) + B(s)u(s), \quad s \geq 0 \tag{3.1}$$

and

$$y'(s) = A(s)y(s) + B(s)u(s), \quad s \geq t; \quad y(t) = x. \tag{3.2}$$

Here $t \geq 0$, $x \in \mathbb{R}^n$; $[A(\cdot), B(\cdot)]$ is a T-periodic pair, i.e.,

$$\begin{aligned} A(\cdot) \in L^1_{loc}(\mathbb{R}^+; \mathbb{R}^{n\times n}), \quad B(\cdot) \in L^\infty(\mathbb{R}^+; \mathbb{R}^{n\times m}); \\ A(t+T) = A(t) \text{ and } B(t+T) = B(t) \text{ for a.e. } t \in \mathbb{R}^+; \end{aligned} \tag{3.3}$$

u is taken respectively from $\mathscr{U}_{ad} \triangleq L^2(0,\infty; \mathbb{R}^m)$ and $\mathscr{U}^t_{ad} \triangleq L^2(t,\infty; \mathbb{R}^m)$. Notice that $\mathscr{U}_{ad} = \mathscr{U}^0_{ad}$. Write $\widehat{\Phi}_A(\cdot)$ for the fundamental solution associated with $A(\cdot)$. Let $\Phi_A(s,t) \triangleq \widehat{\Phi}_A(s)\widehat{\Phi}_A(t)^{-1}$ for all $s \geq t \geq 0$. Then $\Phi_A(\cdot,\cdot)$ is the evolution generated by $A(\cdot)$, which is introduced in Definition 1.3 with $H = \mathbb{R}^n$. Notice that $\Phi_A(\cdot,\cdot)$ is T-periodic (see Proposition 1.2), but $\widehat{\Phi}_A(\cdot)$ is not T-periodic, in general (see, for instance, Example 4.1). Denote by $y(\cdot; t, x, u)$ the solution of Eq. (3.2), and also the

G. Wang and Y. Xu, *Periodic Feedback Stabilization for Linear Periodic Evolution Equations*, SpringerBriefs in Mathematics,
DOI 10.1007/978-3-319-49238-4_3

solution of (3.1) with the initial condition that $y(t) = x$, if there is no risk to cause any confusion. Then we have

$$y(s;t,x,u) = \Phi_A(s,t)x + \int_t^s \Phi_A(s,r)B(r)u(r)\mathrm{d}r \quad \text{for any} \quad s \geq t. \tag{3.4}$$

Let

$$\mathscr{P} \triangleq \mathscr{P}(0) = \widehat{\Phi}_A(T). \tag{3.5}$$

It is exactly the periodic map introduced in Definition 1.4. We will write respectively $\widehat{\Phi}(\cdot)$ and $\Phi(\cdot,\cdot)$ for $\widehat{\Phi}_A(\cdot)$ and $\Phi_A(\cdot,\cdot)$, if it will not cause any confusion.

Definition 3.1 Let $[A(\cdot), B(\cdot)]$ be a T-periodic pair and let $k \in \mathbb{N}$. Equation (3.1) (or the pair $[A(\cdot), B(\cdot)]$) is said to be kT-periodically stabilizable if there exists a kT-periodic $K(\cdot)$ in $L^\infty(\mathbb{R}^+;\mathbb{R}^{m\times n})$ so that the following equation is exponentially stable:

$$\dot{y}(s) = \big[A(s) + B(s)K(s)\big]y(s), \quad s \geq 0. \tag{3.6}$$

Any such a $K(\cdot)$ is called a kT-periodic feedback stabilization law for the pair $\big[A(\cdot), B(\cdot)\big]$ (or for Eq. (3.1)).

3.1 Null Controllable Subspaces

In the finite dimensional case, we will use the null-controllable subspace to play the role of the attainable subspace in the infinite dimensional case. The reason is explained in Remark 3.2.

Definition 3.2 The null-controllable subspace w.r.t. a T-periodic pair $[A(\cdot), B(\cdot)]$ is

$$V_{[A(\cdot),B(\cdot)]} \triangleq \left\{ x \in \mathbb{R}^n \;\middle|\; \exists\, u \in \mathscr{U}_{ad} \text{ and } s > 0 \text{ s.t. } y(s;0,x,u) = 0 \right\}. \tag{3.7}$$

For each $k \in \mathbb{N}$, the null-controllable subspace w.r.t. $[A(\cdot), B(\cdot)]$, on $[0, kT]$, is

$$V_{[A(\cdot),B(\cdot)],k} \triangleq \left\{ x \in \mathbb{R}^n \;\middle|\; \exists\, u \in \mathscr{U}_{ad} \text{ s.t. } y(kT;0,x,u) = 0 \right\}. \tag{3.8}$$

Simply write V and V_k for $V_{[A(\cdot),B(\cdot)]}$ and $V_{[A(\cdot),B(\cdot)],k}$ respectively, when there is no risk to cause any confusion.

Lemma 3.1 *Let $[A(\cdot), B(\cdot)]$ be a T-periodic pair. Assume that matrices $Q \in \mathbb{R}^{n\times n}$ and $R \in \mathbb{R}^{m\times m}$ satisfy (1.88). Then Problem $(LQ)^\infty_{0,x}$ (defined by (1.89) with $(t,h) = (0,x)$) satisfies the FCC for each $x \in V$.*

Proof Let $x \in V$. Then there is $u \in \mathscr{U}_{ad}$ and $s_0 > 0$ so that $y(s_0; 0, x, u) = 0$. Define $\hat{u}(s) \triangleq \chi_{[0,s_0]}(s)u(s)$, $s \in [0, \infty)$. It is clear that $\hat{u} \in \mathscr{U}_{ad}$ and $y(\cdot; 0, x, \hat{u})\big|_{(s_0,\infty)} \equiv 0$. From these, we find that

$$
\begin{aligned}
J_{0,x}^{\infty}(\hat{u}) &= \int_0^{\infty} \Big[\langle y(s; 0, x, \hat{u}),\ Qy(s; 0, x, \hat{u})\rangle + \langle \hat{u}(s),\ R\hat{u}(s)\rangle\Big] ds \\
&= \int_0^{s_0} \Big[\langle y(s; 0, x, \hat{u}),\ Qy(s; 0, x, \hat{u})\rangle + \langle \hat{u}(s),\ R\hat{u}(s)\rangle\Big] ds < \infty.
\end{aligned}
$$

This ends the proof. □

Remark 3.1 By Theorem 1.4, we see that Eq. (3.1) is linear periodic feedback stabilizable if and only if there are positive definite matrices Q and R so that the corresponding LQ problem $(LQ)_{0,x}^{\infty}$ satisfies the FCC for any $x \in \mathbb{R}^n$. On the other hand, from Lemma 3.1, we find that for any positive definite matrices Q and R, the corresponding LQ problem $(LQ)_{0,x}^{\infty}$ satisfies the FCC for any x in $V_{[A(\cdot),B(\cdot)]}$. Thus, Eq. (3.1) is linear periodic feedback stabilizable if and only if there are positive definite matrices Q and R so that the corresponding LQ problem $(LQ)_{0,x}^{\infty}$ satisfies the FCC for any $x \in V_{[A(\cdot),B(\cdot)]}^{\perp}$. The key now is to give a condition ensuring the existence of positive definite matrices Q and R so that the corresponding LQ problem $(LQ)_{0,x}^{\infty}$ satisfies the FCC for any $x \in V_{[A(\cdot),B(\cdot)]}^{\perp}$.

We next introduce some properties on controllable subspaces.

Lemma 3.2 *Let $\big[A(\cdot), B(\cdot)\big]$ be a T-periodic pair with $\mathscr{P}$, V and V_k, which are given by (3.5), (3.7) and (3.8), respectively. Then*

$$
V_k = \sum_{j=0}^{k-1} \mathscr{P}^{-j} V_1 \ \text{ for all } \ k \in \mathbb{N}; \qquad V = V_n. \tag{3.9}
$$

Moreover, it holds that

$$
\mathscr{P}V = V = \mathscr{P}^{-1}V \ \text{ and } \ \mathscr{P}^* V^{\perp} = V^{\perp} = (\mathscr{P}^*)^{-1} V^{\perp}. \tag{3.10}
$$

Proof From (3.8) and (3.4), we find that

$$
V_k = \left\{ \int_0^{kT} \widehat{\Phi}^{-1}(s)B(s)u(s)ds \ \Big| \ u(\cdot) \in \mathscr{U}_{ad} \right\}. \tag{3.11}
$$

We prove (3.9) and (3.10) by three steps as follows:

Step 1. To show the first equality in (3.9)

Arbitrarily fix a $k \in \mathbb{N}$. By the periodicity of $B(\cdot)$, we find that

$$\int_0^{kT} \widehat{\Phi}^{-1}(s)B(s)u(s)\mathrm{d}s = \sum_{j=0}^{k-1} \int_{jT}^{(j+1)T} \widehat{\Phi}^{-1}(s)B(s)u(s)\mathrm{d}s$$
$$= \sum_{j=0}^{k-1} \int_0^T \mathscr{P}^{-j} \widehat{\Phi}^{-1}(s)B(s)u(jT+s)\mathrm{d}s \in \sum_{j=0}^{k-1} \mathscr{P}^{-j} V_1,$$

which leads to $V_k \subseteq \sum_{j=0}^{k-1} \mathscr{P}^{-j} V_1$.

Conversely, let $\hat{x} = \sum_{j=0}^{k-1} \mathscr{P}^{-j} x_{j+1}$, with $x_1, \ldots, x_k \in V_1$. Then there are $u_1(\cdot), \ldots, u_k(\cdot)$ in $\mathscr{U}_{ad}$ so that

$$x_j = \int_0^T \widehat{\Phi}^{-1}(s)B(s)u_j(s)\mathrm{d}s \quad \text{for all} \quad j = 1, \ldots, k.$$

Define a control $\hat{u} \in \mathscr{U}_{ad}$ by setting $\hat{u}(jT+s) = u_j(s)$ for all $j \in \{0, 1, \ldots, k-1\}$, $s \in [0, T)$ and $\hat{u}(s) = 0$ for all $s \in [kT, \infty)$. Then, we see that

$$\hat{x} = \int_0^{kT} \widehat{\Phi}^{-1}(s)B(s)\hat{u}(s)\mathrm{d}s,$$

which leads to $\hat{x} \in V_k$. From this, we find that $\sum_{j=0}^{k-1} \mathscr{P}^{-j} V_1 \subseteq V_k$, which shows the first equality in (3.9).

Step 2. To prove the second equality in (3.9)

It is obvious that $V_n \subseteq V$. To show the reverse, let $x \in V$. By (3.7), there is $u(\cdot) \in \mathscr{U}_{ad}$ and $s_0 > 0$ so that $y(s_0; 0, x, u) = 0$. Let $\hat{u}(s) = \chi_{(0,s_0)}(s)u(s)$ for $s \geq 0$ and $N(s_0)$ be the integer so that $N(s_0)T < s_0 \leq (N(s_0)+1)T$. Then

$$y((N(s_0)+1)T; 0, x, \hat{u}) = y((N(s_0)+1)T; s_0, 0, \hat{u}) = 0,$$

from which, it follows that $x \in V_{N(s_0)+1}$. On the other hand, by the Hamilton-Cayley theorem, each $\mathscr{P}^{-j} V_1$, with $j \geq n$, is a linear combination of V_1, $\mathscr{P}^{-1} V_1, \ldots$, $\mathscr{P}^{-(n-1)} V_1$. This, together with the first equality in (3.9), indicates that

$$V_{N(s_0)+1} = \sum_{j=0}^{N(s_0)} \mathscr{P}^{-j} V_1 \subseteq \sum_{j=0}^{n-1} \mathscr{P}^{-j} V_1 = V_n.$$

Therefore, $x \in V_n$ which leads to $V = V_n$.

Step 3. To show (3.10)
It is clear that

$$\mathscr{P}^{-1}V = \mathscr{P}^{-1}V_n = \mathscr{P}^{-1}\sum_{j=0}^{n-1}\mathscr{P}^{-j}V_1 = \sum_{j=1}^{n}\mathscr{P}^{-j}V_1 \subseteq \sum_{j=0}^{n-1}\mathscr{P}^{-j}V_1 = V_n.$$

This, along with the second equality in (3.9), indicates that $\mathscr{P}^{-1}V \subseteq V$. Meanwhile, because $\mathscr{P}$ is invertible, $\dim(\mathscr{P}^{-1}V) = \dim V$. Thus, we find that $\mathscr{P}^{-1}V = V$, i.e., $V = \mathscr{P}V$. Hence, the first statement in (3.10) stands. The second statement in (3.10) is a direct consequence of the first one. This completes the proof. □

We next provide an algebraic characterization of $V = V_n$ via the LQ theory. Consider the following equation:

$$\begin{cases} \dot{\Upsilon}(t) + A(t)^*\Upsilon(t) + \Upsilon(t)A(t) - \frac{1}{\varepsilon}\Upsilon(t)B(t)B(t)^*\Upsilon(t) = 0, \ t \in [0, nT]; \\ \Upsilon(nT) = I. \end{cases} \tag{3.12}$$

It is indeed the Riccati Eq. (1.61) with $Q = 0$, $R = \varepsilon I$, $M = I$ and $\hat{T} = nT$. By Theorem 1.1, Eq. (3.12) admits a unique solution $P_n^{\varepsilon}(\cdot) \in C_u([0, nT]; \mathscr{SL}(\mathbb{R}^n))$, with $P_n^{\varepsilon}(t) \geq 0$ for each $t \in [0, nT]$. Because $\mathscr{SL}(\mathbb{R}^n)$ is of finite dimension, the spaces $C_u([0, nT]; S\mathscr{L}(\mathbb{R}^n))$ and $C([0, nT]; S\mathscr{L}(\mathbb{R}^n))$ are the same. We further find that for each $t \in [0, nT]$, $P_n^{\varepsilon}(t) \gg 0$ (i.e., $P_n^{\varepsilon}(t)$ is positive definite). Indeed, since we are working in a finite dimensional space, it suffices to show that $\langle x, P_n^{\varepsilon}(t)x\rangle > 0$ for all $t \in [0, nT)$ and $x \in \mathbb{R} \setminus \{0\}$. By contradiction, suppose that there was $t_0 \in [0, nT)$ and $x_0 \in \mathbb{R} \setminus \{0\}$ so that $\langle x_0, P_n^{\varepsilon}(t)x_0\rangle = 0$. From (i) of Theorem 1.2, the optimal trajectory $\bar{y}(\cdot)$ to $(LQ)_{t_0,x_0}^{nT}$ solves Eq. (1.2) with $h = x_0$ and $\hat{T} = nT$. Hence, $\bar{y}(nT) \neq 0$. So we have that $J_{t_0,x_0}^{nT,\varepsilon}(\bar{u}) \geq \langle \bar{y}(nT), \bar{y}(nT)\rangle > 0$, where $\bar{u}$ is the corresponding optimal control. On the other hand, by (ii) of Theorem 1.2, we see that

$$W_{t_0,x_0}^{nT,\varepsilon} = J_{t_0,x_0}^{nT,\varepsilon}(\bar{u}) = \langle x_0, P_n^{\varepsilon}(t_0)x_0\rangle = 0.$$

These lead to a contradiction.

We now can set

$$S_n^{\varepsilon}(t) = \big(P_n^{\varepsilon}(t)\big)^{-1}, \quad t \in [0, nT]. \tag{3.13}$$

Clearly, $S_n^{\varepsilon}(\cdot)$ solves the following linear equation:

$$\dot{S}_n(t) - A(t)S_n(t) - S_n(t)A(t)^* + \frac{1}{\varepsilon}B(t)B(t)^* = 0, \quad t \in [0, nT]; \quad S_n(nT) = I. \tag{3.14}$$

Arbitrarily fix $x \in \mathbb{R}^n$ and $u(\cdot) \in L^2(0, nT; \mathbb{R}^m)$. One can easily check that $J_{0,x}^{nT,\varepsilon_1}(u) \leq J_{0,x}^{nT,\varepsilon_2}(u)$ for all $\varepsilon_1 \leq \varepsilon_2$. Thus, $\langle x, P_n^{\varepsilon}(0)x\rangle$ (which equals to $W^{nT,\varepsilon}(0, x)$ by Theorem 1.2) is monotonically increasing with respect to ε and bounded from below by 0. Hence, $\lim_{\varepsilon\to 0+}\langle x, P_n^{\varepsilon}(0)x\rangle$ exists for each $x \in \mathbb{R}^n$. Define a bilinear function $l(\cdot,\cdot):\mathbb{R}^n\times\mathbb{R}^n\mapsto\mathbb{R}$ by $l(x,y)\triangleq\lim_{\varepsilon\to0+}\langle x,\ P_n^\varepsilon(0)y\rangle$. Then we find that for all $x,y\in\mathbb{R}^n$, $l(x,y)=l(y,x)$, $l(x,x)\ge0$ and

$$l(x,y)=\frac{1}{2}\lim_{\varepsilon\to0+}\left[\langle(x+y),\ P_n^\varepsilon(0)(x+y)\rangle-\langle x,\ P_n^\varepsilon(0)x\rangle-\langle y,\ P_n^\varepsilon(0)y\rangle\right].$$

Thus, there is a positive semi-definite matrix $\bar{Q}$ so that $l(x,y)=\langle x,\ \bar{Q}y\rangle$ for all $x,y\in\mathbb{R}^n$. Hence, we find that for all $x,y\in\mathbb{R}^n$, $\langle x,\ \bar{Q}y\rangle=\lim_{\varepsilon\to0+}\langle x,\ P_n^\varepsilon(0)y\rangle$. From this, it follows that

$$\bar{Q}=\lim_{\varepsilon\to0+}P_n^\varepsilon(0)\quad\text{in }\mathscr{L}(\mathbb{R}^n).\tag{3.15}$$

The above-mentioned algebraic characterization of V is presented in the following Lemma.

Lemma 3.3 *Let $[A(\cdot),\ B(\cdot)]$ be a T-periodic pair with V and $\bar{Q}$ given by (3.7) and (3.15) respectively. Then*

$$V=\mathscr{N}(\bar{Q})\quad\text{and}\quad V^\perp=\mathscr{R}(\bar{Q}).\tag{3.16}$$

Proof We first claim that $x\in V$ if and only if $\langle x,\ \bar{Q}x\rangle=0$. In fact, on one hand, if $x\in V$, then it follows from Lemma 3.2 that $x\in V_n$, i.e., there is a control $\hat{u}(\cdot)\in\mathscr{U}_{ad}$ so that $y(nT;0,x,\hat{u})=0$. We restrict this control on $[0,nT]$ and still denote by $\hat{u}(\cdot)$ the restricted control, which is clearly in $L^2(0,nT;\mathbb{R}^m)$. Then, one can easily check that

$$0\le\langle x,\ \bar{Q}x\rangle=\lim_{\varepsilon\to0+}W^{nT,\varepsilon}(0,x)\le\lim_{\varepsilon\to0+}J_{0,x}^{nT,\varepsilon}(\hat{u})=\lim_{\varepsilon\to0+}\varepsilon\int_0^{nT}\|\hat{u}(t)\|^2\mathrm{d}t=0.$$

On the other hand, if $x\in\mathbb{R}^n$ satisfies $\langle x,\ \bar{Q}x\rangle=0$, then by the definition of $J_{0,x}^{nT,\varepsilon}$,

$$0=\langle x,\ \bar{Q}x\rangle=\lim_{\varepsilon\to0+}\inf_{u\in L^2(0,nT;\mathbb{R}^m)}J_{0,x}^{nT,\varepsilon}(u)\ge\inf_{u\in L^2(0,nT;\mathbb{R}^m)}\|y(nT;0,x,u)\|^2.\tag{3.17}$$

It follows by (3.4) that

$$\{y(nT;0,x,u)\mid u\in L^2(0,nT;\mathbb{R}^m)\}=\widehat{\Phi}(nT)x+\widehat{\Phi}(nT)V.$$

This, along with (3.17), yields that

$$\inf_{z\in\widehat{\Phi}(nT)x+\widehat{\Phi}(nT)V}\|z\|^2=0.$$

Because $\widehat{\Phi}(nT)x+\widehat{\Phi}(nT)V$ is closed, we find that $0\in\widehat{\Phi}(nT)x+\widehat{\Phi}(nT)V$, which leads to $x\in V$. Hence, we have proved the above claim. Now, since $\bar{Q}$ is symmetric and positive semi-definite, $\langle x,\ \bar{Q}x\rangle=0$ if and only if $\bar{Q}x=0$. This, along with the above claim, leads to the first equality in (3.16). The second equality in (3.16) follows from the following facts:

$$V^{\perp} = (\mathscr{N}(\bar{Q}))^{\perp} = \mathscr{R}(\bar{Q}^*) = \mathscr{R}(\bar{Q}).$$

This ends the proof. □

The next lemma is concerned with properties of a decomposition of the solution $y(\cdot; 0, x, u)$ to Eq. (3.2) with $t = 0$.

Lemma 3.4 *Let* $\big[A(\cdot), B(\cdot)\big]$ *be a* T*-periodic pair with* $\mathscr{P}$, V *and* $\bar{Q}$ *is given by (3.5), (3.7) and (3.15). Given* $x \in \mathbb{R}^n$, $u \in \mathscr{U}_{ad}$ *and* $s \geq 0$, *write*

$$y_1(s; 0, x, u) \triangleq \mathrm{Proj}_{\widehat{\Phi}(s)V}\, y(s; 0, x, u) \;\; and \;\; y_2(s; 0, x, u) \triangleq \mathrm{Proj}_{(\widehat{\Phi}(s)V)^{\perp}}\, y(s; 0, x, u). \tag{3.18}$$

Then

$$y_2(s; 0, x, u) = \mathrm{Proj}_{(\widehat{\Phi}(s)V)^{\perp}} \widehat{\Phi}(s) \mathrm{Proj}_{V^{\perp}} x \;\; for\ each \;\; s \geq 0. \tag{3.19}$$

where $x_2 = \mathrm{Proj}_{V^{\perp}} x$. *In particular, it holds that*

$$y_2(nT; 0, x, u) = \bar{Q}^{\dagger} \bar{Q} \mathscr{P}^n x_2 \in V^{\perp}. \tag{3.20}$$

Proof Let $x \in \mathbb{R}^n$ and $u \in \mathscr{U}_{ad}$. Write $x_1 \triangleq \mathrm{Proj}_V x$ and $x_2 \triangleq \mathrm{Proj}_{V^{\perp}} x$. Since

$$V = \left\{ \int_0^s \widehat{\Phi}^{-1}(r) B(r) \hat{u}(r) \mathrm{d}r \;\middle|\; \hat{u}(\cdot) \in \mathscr{U}_{ad},\ s \in \mathbb{R}^+ \right\},$$

we have that for each $s \geq 0$,

$$\widehat{\Phi}(s) \left(x_1 + \int_0^s \widehat{\Phi}(r)^{-1} B(r) u(r) \mathrm{d}r \right) \in \widehat{\Phi}(s) V.$$

This, along with (3.4), yields that for each $s \geq 0$,

$$\begin{aligned} y(s; 0, x, u) &= \widehat{\Phi}(s)(x_1 + x_2) + \widehat{\Phi}(s) \int_0^s \widehat{\Phi}(r)^{-1} B(r) u(r) \mathrm{d}r \\ &= \widehat{\Phi}(s) \left(x_1 + \int_0^s \widehat{\Phi}(r)^{-1} B(r) u(r) \mathrm{d}r \right) + \widehat{\Phi}(s) x_2 \subseteq \widehat{\Phi}(s) V + \widehat{\Phi}(s) x_2. \end{aligned} \tag{3.21}$$

Let $z_1(s) = \mathrm{Proj}_{\widehat{\Phi}(s)V} \widehat{\Phi}(s) x_2$ and $z_2(s) = \mathrm{Proj}_{(\widehat{\Phi}(s)V)^{\perp}} \widehat{\Phi}(s) x_2$, with $s \geq 0$. Then, by (3.18) and (3.21), we find that for each $s \geq 0$,

$$y_1(s; 0, x, u) = \widehat{\Phi}(s) \left(x_1 + \int_0^s \widehat{\Phi}(r)^{-1} B(r) u(r) \mathrm{d}r \right) + z_1(s); \;\; y_2(s; 0, x, u) = z_2(s),$$

which leads to (3.19).

Finally, when $s = nT$, it follows from Lemma 3.2 that $(\widehat{\Phi}(nT)V)^{\perp} = V^{\perp}$. This, together with (3.19), shows that $y_2(nT; 0, x, u) = \text{Proj}_{V^{\perp}} \mathscr{P}^n x_2$. Meanwhile, by Lemma 1.1, with $M = \bar{Q}$ and $\xi = \mathscr{P}^n x_2$, and by Lemma 3.3, we find that

$$\bar{Q}^{\dagger} \bar{Q} \mathscr{P}^n x_2 \in \mathscr{R}(\bar{Q}) = V^{\perp}, \qquad \mathscr{P}^n x_2 - \bar{Q}^{\dagger} \bar{Q} \mathscr{P}^n x_2 \in \mathscr{N}(\bar{Q}) = V.$$

Combining these, we get (3.20). This ends the proof. □

Remark 3.2 In our finitely dimensional studies, we will use Lemma 3.4 to replace the decomposition provided by Kato projection. It deserves to mention that the decomposition in this lemma is about solutions with controls u, while the Kato projection provides a decomposition for solutions with the null control. From this point of view, the decomposition in Lemma 3.4 is better than that given by the Kato projection. However, this decomposition works because of the fact that $V = V_n$, which does not seem right in infinitely dimensional cases.

We end this section with introducing some notations which will be used later. Given a T-periodic pair $[A(\cdot), B(\cdot)]$, with V and $\mathscr{P}$. By (3.10), V is invariant with respect to $\mathscr{P}$. Hence, Lemma 1.2 (see also Remark 1.2) provides a unique decomposition $(V_1(\mathscr{P}), V_2(\mathscr{P}))$. Write

$$k_1 \stackrel{\Delta}{=} \dim V, \qquad k_2 \stackrel{\Delta}{=} \dim V_1(\mathscr{P}), \qquad k_3 \stackrel{\Delta}{=} \dim \mathbb{R}_1^n(\mathscr{P}). \tag{3.22}$$

Clearly,

$$k_1 \geq k_2 \leq k_3, \tag{3.23}$$

$$k_1 - k_2 = \dim V_2(\mathscr{P}), \qquad n - k_3 = \dim \mathbb{R}_2^n(\mathscr{P}). \tag{3.24}$$

Since $V_2(\mathscr{P}) \subseteq \mathbb{R}_2^n(\mathscr{P})$ (see Lemma 1.3), we have that

$$k_1 + k_3 \leq n + k_2. \tag{3.25}$$

3.2 Algebraic Criterion and Application

In this section, we present an algebraic criterion of the periodic feedback stabilization. In the first two subsections, we provide its proof, while in the last subsection, we give an application.

For each T-periodic pair $\big[A(\cdot), B(\cdot)\big]$ and $\varepsilon > 0$, two linear ordinary differential equations will be considered. The first one is (3.14), while the second one is as follows:

$$\dot{S}(t) - A(t)S(t) - S(t)A(t)^* + \frac{1}{\varepsilon}B(t)B(t)^* = 0, \quad t \in [0, T], \qquad S(T) = \mathscr{P}XX^*\mathscr{P}^*, \tag{3.26}$$

where X is an invertible matrix in $\mathbb{R}^{n\times n}$. Write $S_n^\varepsilon(\cdot)$ and $S^\varepsilon(\cdot)$ for the unique solutions of Eqs. (3.14) and (3.26) respectively. We have proved in Sect. 3.1 that $S_n^\varepsilon(\cdot)$ is a symmetric and positive definite matrix-valued functions over $[0, nT]$. Similarly, we can show that $S^\varepsilon(\cdot)$ is a symmetric and positive definite matrix-valued functions over $[0, T]$. Thus, for each $\varepsilon > 0$, we can define respectively an nT-periodic matrix-valued function $K_n^\varepsilon(\cdot)$ in $L^\infty(\mathbb{R}^+; \mathbb{R}^{m\times n})$ and a T-periodic matrix-valued function $K^\varepsilon(\cdot)$ in $L^\infty(\mathbb{R}^+; \mathbb{R}^{m\times n})$ by setting

$$K_n^\varepsilon(s) = -\frac{1}{\varepsilon}B^*(s)\big(S_n^\varepsilon(s)\big)^{-1} \text{ a.e. } s\in[0,nT);\quad K_n^\varepsilon(s) = K_n^\varepsilon(s+nT) \text{ a.e. } s>0 \tag{3.27}$$

and

$$K^\varepsilon(s) = -\frac{1}{\varepsilon}B^*(s)\big(S^\varepsilon(s)\big)^{-1} \text{ a.e. } s\in[0,T);\quad K^\varepsilon(s) = K^\varepsilon(s+T) \text{ a.e. } s>0. \tag{3.28}$$

The above-mentioned algebraic criterion is presented as follows:

Theorem 3.1 *Let $\big[A(\cdot), B(\cdot)\big]$ be a T-periodic pair with $\mathscr{P}$ and $\bar{Q}$ given by (3.5) and (3.15) respectively. Then, the following statements are equivalent: (a) $\big[A(\cdot), B(\cdot)\big]$ is nT-periodically stabilizable; (b) $\big[A(\cdot), B(\cdot)\big]$ is T-periodically stabilizable; (c) $\sigma\left(\bar{Q}^\dagger\bar{Q}\mathscr{P}\right)\subset\mathbb{B}$. Furthermore, when $\big[A(\cdot), B(\cdot)\big]$ is T-periodically stabilizable, each $K_n^\varepsilon(\cdot)$ defined by (3.27), with $\left\|\big(S_n^\varepsilon(0)\big)^{-1}-\bar{Q}\right\| < 1$, is an nT-periodic feedback stabilization law for this pair; and meanwhile there is an invertible $X\in\mathbb{R}^{n\times n}$ (depending on $V_{[A(\cdot),B(\cdot)],k}$, with $k=1,\dots n$) and a positive number ε_0 (depending on n, $\|X\|$ and $\|\mathscr{P}\|$) so that each $K^\varepsilon(\cdot)$ given by (3.28), with $\varepsilon\le\varepsilon_0$, is a T-periodic feedback stabilization law for this pair.*

3.2.1 *The Proof of (a)⇔(c) in Theorem* 3.1

It is hidden in the proof of this part that if (c) stands, then $K_n^\varepsilon(\cdot)$ given by (3.27) is an nT-periodic feedback law when ε is sufficiently small. The strategy to prove $(a)\Leftrightarrow(c)$ in Theorem 3.1 is to show that $(a)\Leftrightarrow\sigma\left(\bar{Q}^\dagger\bar{Q}\mathscr{P}^n\right)\subset\mathbb{B}$ and $(c)\Leftrightarrow\sigma\left(\bar{Q}^\dagger\bar{Q}\mathscr{P}^n\right)\subset\mathbb{B}$. Their proof will be carried by three steps as follows:

Step 1: To prove $(a)\Rightarrow\sigma\left(\bar{Q}^\dagger\bar{Q}\mathscr{P}^n\right)\subset\mathbb{B}$

Let $(y_1(\cdot;0,x,u), y_2(\cdot;0,x,u))$ be the decomposition of $y(\cdot;0,x,u)$ provided by Lemma 3.4. Then it follows from (3.19) that

$$y_2(s;0,x,u) = y_2(s;0,x,0),\quad \text{when } s\ge 0, u\in\mathscr{U}_{ad}. \tag{3.29}$$

We claim that

$$\sigma\left(\bar{Q}^\dagger\bar{Q}\mathscr{P}^n\right)\subset\mathbb{B}\Leftrightarrow\lim_{s\to\infty}y_2(s;0,x,0)=0 \text{ for any } x\in\mathbb{R}^n. \tag{3.30}$$

To this end, it suffices to show that

$$\lim_{s\to\infty} y_2(s;0,x,0)=0 \ \text{ for any } x\in\mathbb{R}^n \Leftrightarrow \lim_{k\to\infty}\left(\bar{Q}^\dagger\bar{Q}\mathscr{P}^n\right)^k z=0 \ \text{ for any } z\in V^\perp, \tag{3.31}$$

and

$$\lim_{k\to\infty}\left(\bar{Q}^\dagger\bar{Q}\mathscr{P}^n\right)^k z=0 \ \text{ for any } \ z\in V^\perp \Leftrightarrow \sigma\left(\bar{Q}^\dagger\bar{Q}\mathscr{P}^n\right)\subset\mathbb{B}. \tag{3.32}$$

To prove (3.31), we first show that for any $x\in\mathbb{R}^n$,

$$\lim_{s\to\infty} y_2(s;0,x,0)=0 \Leftrightarrow \lim_{k\to\infty} y_2(knT;0,x,0)=0. \tag{3.33}$$

Indeed, we only need to show the right hand side of (3.33) implies the left hand side of (3.33), since the reverse is obvious. To this end, we let, for each $s\geq 0$, $N(s)$ be the non-negative integer so that $N(s)nT<s\leq(N(s)+1)nT$. Let $\hat{y}(s)\triangleq y(N(s)nT;0,x,0)$ and $\hat{s}=s-N(s)nT$ for each $s\geq 0$. By the T-periodicity of the system, we find that

$$y(s;0,x,0)=y(\hat{s};0,\hat{y}(s),0) \ \text{ and } \ \widehat{\Phi}(s)V=\widehat{\Phi}(\hat{s})\mathscr{P}^{N(s)n}V=\widehat{\Phi}(\hat{s})V \quad \text{for all } \ s\geq 0.$$

Thus we have that

$$\begin{aligned} y_2(s;0,x,0) &= \mathrm{Proj}_{(\widehat{\Phi}(s)V)^\perp} y(s;0,x,0) \\ &= \mathrm{Proj}_{(\widehat{\Phi}(\hat{s})V)^\perp} y(\hat{s};0,\hat{y}(s),0)=y_2(\hat{s};0,\hat{y}(s),0), \ \ s\geq 0. \end{aligned} \tag{3.34}$$

Let $\hat{x}_2(s)\triangleq \mathrm{Proj}_{V^\perp}\hat{y}(s)$ for each $s\geq 0$. Because

$$\mathrm{Proj}_{V^\perp}\hat{y}(s)=\mathrm{Proj}_{(\widehat{\Phi}(N(s)nT))V^\perp} y(N(s)nT;0,x,0)=y_2(N(s)nT;0,x,0), \quad s\geq 0,$$

we see that

$$\hat{x}_2(s)=y_2(N(s)nT;0,x,0), \quad s\geq 0. \tag{3.35}$$

Meanwhile, from (3.19), we find that

$$y_2(\hat{s};0,\hat{y}(s),0)=\mathrm{Proj}_{(\widehat{\Phi}(\hat{s})V)^\perp}(\widehat{\Phi}(\hat{s})\hat{x}_2(s)), \quad s\geq 0. \tag{3.36}$$

It follows from (3.34), (3.35) and (3.36) that for each $s\geq 0$,

$$\begin{aligned} &\|y_2(s;0,x,0)\|=\|y_2(\hat{s};0,\hat{y}(s),0)\|=\|\mathrm{Proj}_{(\widehat{\Phi}(\hat{s})V)^\perp}(\widehat{\Phi}(\hat{s})\hat{x}_2(s))\|\leq\|\widehat{\Phi}(\hat{s})\hat{x}_2(s)\| \\ &\leq \max_{r\in[0,nT]}\|\widehat{\Phi}(r)\|\cdot\|\hat{x}_2(s)\|=\max_{r\in[0,nT]}\|\widehat{\Phi}(r)\|\cdot\|y_2(N(s)nT;0,x;0)\|. \end{aligned}$$

This yields (3.33).

On the other hand, one can easily derive from (3.20) that

$$y_2(knT; 0, x; 0) = \left(\bar{Q}^\dagger \bar{Q} \mathscr{P}^n\right)^k \mathrm{Proj}_{V^\perp} x \in V^\perp \ \text{ for all } \ k \in \mathbb{N} \ \text{ and } \ x \in \mathbb{R}^n. \tag{3.37}$$

Now (3.31) follows from (3.33) and (3.37).

We next verify (3.32). It is well known that the right hand side of (3.32) implies the left hand side of (3.32) (see Appendix C5 in [86]). Now we show the reverse. From Lemma 1.1,

$$\mathscr{P}^n z - \bar{Q}^\dagger \bar{Q} \mathscr{P}^n z \in \mathscr{N}(\bar{Q}) \ \text{ for each } \ z \in \mathbb{R}^n. \tag{3.38}$$

Meanwhile, it follows by Lemmas 3.2 and 3.3 that

$$z \in V \Rightarrow \mathscr{P}^n z \in V = \mathscr{N}(\bar{Q}). \tag{3.39}$$

By (3.38) and (3.39), we find that $\bar{Q}^\dagger \bar{Q} \mathscr{P}^n z = 0$ for all $z \in V$. From this and the left hand side of (3.32), it follows that

$$\lim_{k \to \infty} \left(\bar{Q}^\dagger \bar{Q} \mathscr{P}^n\right)^k z = 0 \ \text{ for all } \ z \in \mathbb{R}^n.$$

This yields the right hand side of (3.32) (see Appendix C5 in [86]). Hence, the claim (3.30) has been proved.

We now suppose that (a) in Theorem 3.1 stands. Let $K(\cdot)$ be an nT-periodic stabilization law for the pair $\left[A(\cdot), B(\cdot)\right]$. Consider the following equation:

$$\dot{y}(s) = \left[A(s) + B(s)K(s)\right] y(s), \quad s \in \mathbb{R}^+; \quad y(0) = x. \tag{3.40}$$

For each $x \in \mathbb{R}^n$, we denoted by $y^K(\cdot; x)$ the unique solution of Eq. (3.40). Since $K(\cdot)$ is a feedback stabilization law, we see that

$$\lim_{s \to \infty} y^K(s; x) = 0 \ \text{ for each } \ x \in \mathbb{R}^n. \tag{3.41}$$

Given $x \in \mathbb{R}^n$, we let $u^x(\cdot) \triangleq K(\cdot) y^K(\cdot; x)$. It is clear that

$$y^K(\cdot; x) = y(\cdot; 0, x, u^x). \tag{3.42}$$

Denote by $(y_1(\cdot; 0, x, u^x), y_2(\cdot; 0, x, u^x))$ the decomposition of the solution $y(\cdot; 0, x, u^x)$ provided by Lemma 3.4. From the orthogonality of the decomposition, it follows that

$$\|y_2(s; 0, x, u^x))\| \le \sqrt{\|y_1(s; 0, x, u^x))\|^2 + \|y_2(s; 0, x, u^x))\|^2} = \|y^K(s; x)\| \ \text{ for all } \ s \ge 0.$$

This, along with (3.41), yields that

$$\lim_{s\to\infty} y_2(s;0,x,u^x)) = 0 \ \text{ for each } \ x\in\mathbb{R}^n. \tag{3.43}$$

In summary, we conclude from (3.43), (3.29) and (3.30) that $\sigma\left(\bar{Q}^\dagger\bar{Q}\mathscr{P}^n\right)\subset\mathbb{B}$.

Step 2: To show $\sigma\left(\bar{Q}^\dagger\bar{Q}\mathscr{P}^n\right)\subset\mathbb{B}\Rightarrow(a)$

Suppose that $\sigma\left(\bar{Q}^\dagger\bar{Q}\mathscr{P}^n\right)\subset\mathbb{B}$. It follows from (3.15) and (3.13) that there is an $\varepsilon_0>0$ so that $\|P_n^\varepsilon(0)-\bar{Q}\|<1$, when $0<\varepsilon\le\varepsilon_0$. We arbitrarily fix an $\varepsilon\in(0,\varepsilon_0]$, and then write

$$\Lambda_1 \triangleq \Lambda_1(\varepsilon) \triangleq \|Q_n^\varepsilon(0)-\bar{Q}\|^{1/2} = \|(S_n^\varepsilon(0))^{-1}-\bar{Q}\|^{\frac{1}{2}} < 1. \tag{3.44}$$

Let $K_n^\varepsilon(\cdot)$ be given by (3.27). It suffices to show that $K_n^\varepsilon(\cdot)$ is an nT-periodic stabilization law for $[A(\cdot),B(\cdot)]$. For this purpose, we write $\Psi_\varepsilon(\cdot)$ for the fundamental solution associated with $A(\cdot)+B(\cdot)K_n^\varepsilon(\cdot)$ and write $\mathscr{P}_\varepsilon\triangleq\Psi_\varepsilon(nT)$. Given $t\ge0$ and $x\in\mathbb{R}^n$, let $y^\varepsilon(\cdot;t,x)$ be the unique solution to the equation:

$$\dot{y}(s) = \Big(A(s)+B(s)K_n^\varepsilon(s)\Big)y(s), \quad s\in\mathbb{R}^+; \quad y(t)=x. \tag{3.45}$$

Clearly, $y^\varepsilon(\cdot;t,x)$ is also the unique solution to Eq. (3.1), with the initial condition that $y(t)=x$, and with $u(\cdot)=K_n^\varepsilon(\cdot)y^\varepsilon(\cdot;t_0,x)$. Write $(y_1(\cdot;t,x),y_2(\cdot;t,x))$ for the decomposition of $y^\varepsilon(\cdot;t,x)$ provided by Lemma 3.4 with the control $u(\cdot)=K_n^\varepsilon(\cdot)y^\varepsilon(\cdot;t,x)$. Then the pair $(y_1(\cdot;t,x),y_2(\cdot;t,x))$ satisfies (3.19) and (3.20). The key is to show that

$$\exists\,\bar{k}\in\mathbb{N}, \ \text{ s.t. } \ \lim_{j\to\infty} y^\varepsilon(j\bar{k}nT;0,x)=0 \ \text{ for all } \ x\in\mathbb{R}^n. \tag{3.46}$$

When (3.46) is proved, we have that $\sigma\left(\mathscr{P}_\varepsilon\right)\subset\mathbb{B}$ (see Appendix C5 in [86]). Hence, $K_n^\varepsilon(\cdot)$ is an nT-periodic stabilization law for $[A(\cdot),B(\cdot)]$ (see [71] or [70]).

The rest of this step is to show (3.46). The proof is built upon the following two cases:

In the first case where $x\in V$, we begin with proving that

$$\|y^\varepsilon(nT;0,z)\| \le \Lambda_1\|z\| \ \text{ for all } \ z\in V. \tag{3.47}$$

For this purpose, we observe from (3.16) that

$$\langle z,\bar{Q}z\rangle = 0 \ \text{ for all } \ z\in V. \tag{3.48}$$

Notice that (3.12) is the Riccati equation associated with $\{(LQ)^{nT,\varepsilon}\}_{(t,x)\in[0,nT)\times\mathbb{R}^n}$, where the system is (3.1) and the cost functional is as:

$$\int_t^{nT} \varepsilon \|u(s)\|^2 ds + \|y(nT; t, x, u)\|^2 \text{ for all } u \in L^2(0, nT; \mathbb{R}^m). \tag{3.49}$$

It follows from Theorem 1.2 that the control $\bar{u}_z^\varepsilon(\cdot)$ (whit $z \in V$), defined by

$$\bar{u}_z^\varepsilon(s) \triangleq K_n^\varepsilon(s) y^\varepsilon(s; 0, z) = -\frac{1}{\varepsilon} B^*(s) P_n^\varepsilon(s) y^\varepsilon(s; 0, z) \text{ for a.e. } s \in [0, nT],$$

is the optimal control to Problem $(LQ)_{0,z}^{nT,\varepsilon}$. Furthermore, $y^\varepsilon(\cdot; 0, z)$ is the corresponding optimal trajectory. This, along with (iii) of Theorem 1.2, (3.44) and (3.48), indicates that

$$\|y^\varepsilon(nT; 0, z))\|^2 \le W^{nT,\varepsilon}(0, z) = \langle z, P_n^\varepsilon(0) z\rangle \le \Lambda_1^2 \|z\|^2 + \langle z, \bar{Q} z\rangle = \Lambda_1^2 \|z\|^2,$$

which leads to (3.47). Next, since $x \in V$, it follows from (3.19) that $y_2(\cdot; 0, x) \equiv 0$. From (3.10), we see that

$$y^\varepsilon(knT; 0, x) = y_1(knT; 0, x) \in \widehat{\Phi}(knT) V = \mathscr{P}^{kn} V = V \quad \text{for all } k \in \mathbb{N}.$$

Let $z \triangleq y^\varepsilon(knT; 0, x)$. By the nT-periodicity of $\big(A(\cdot) + B(\cdot) K_n^\varepsilon(\cdot)\big)$, we see that

$$y^\varepsilon(nT; 0, z) = y^\varepsilon((k+1)nT; 0, x).$$

Thus, it follows from (3.47) that

$$\|y^\varepsilon((k+1)nT; 0, x)\| \le \Lambda_1 \|y^\varepsilon(knT; 0, x)\| \text{ for all } k \in \mathbb{N} \text{ and } x \in V. \tag{3.50}$$

Since $\Lambda_1 < 1$ (see (3.44)), it holds that

$$\lim_{k \to \infty} y^\varepsilon(knT; 0, x) = 0 \text{ for all } x \in V. \tag{3.51}$$

In the second case when $x \in V^\perp$, since $\sigma\left(\bar{Q}^\dagger \bar{Q} \mathscr{P}^n\right) \subset \mathbb{B}$, there is a natural number $\bar{k}$ so that (see Appendix C5 in [86])

$$\Lambda_2 \triangleq \|(\bar{Q}^\dagger \bar{Q} \mathscr{P}^n)^{\bar{k}}\| < 1. \tag{3.52}$$

Let

$$a_j \triangleq y_1(j\bar{k}nT; 0, x) \text{ and } \beta_j \triangleq y_2(j\bar{k}nT; 0, x), \ j = 0, 1, 2, \ldots. \tag{3.53}$$

Clear, when $x \in V^\perp$, $\alpha_j + \beta_j = y^\varepsilon(j\bar{k}nT; 0, x)$ for all $j \in \mathbb{N}$; $\alpha_0 = 0$ and $\beta_0 = x$. By the nT-periodicity of $\big(A(\cdot) + B(\cdot) K_n^\varepsilon(\cdot)\big)$, one can easily check that

$$\alpha_{j+1} = y^\varepsilon\left(\bar{k}nT; 0, \alpha_j\right) + y^\varepsilon\left(\bar{k}nT; 0, \beta_j\right) - \beta_{j+1}. \tag{3.54}$$

Because $\{\alpha_j\}_{j=0}^{\infty} \subset V$ (see (3.18) and (3.10)), it follows from (3.50) that

$$\|y^{\varepsilon}(\bar{k}nT; 0, \alpha_j)\| \le \Lambda_1^{\bar{k}} \|\alpha_j\| \quad \text{for all} \quad j = 0, 1, 2, \ldots. \tag{3.55}$$

Recall that $\Psi_{\varepsilon}(\cdot)$ is the fundamental solution associated with $\big(A(\cdot) + B(\cdot)K_n^{\varepsilon}(\cdot)\big)$. Clearly,

$$\left\| y^{\varepsilon}\left(\bar{k}nT; 0, \beta_j\right) \right\| \le \left\| \Psi_{\varepsilon}\left(\bar{k}nT\right) \right\| \cdot \left\| \beta_j \right\|. \tag{3.56}$$

Meanwhile, by (3.20) (see Lemma 3.4) and the nT-periodicity of (3.1), we can easily check that $y_2(knT; 0, x) = \left(\bar{Q}^{\dagger}\bar{Q}\mathscr{P}^n\right)^k x$ for any $k \in \mathbb{N}$ and that $\beta_{j+1} = \left(\bar{Q}^{\dagger}\bar{Q}\mathscr{P}^n\right)\beta_j$ for any $j = 0, 1, \ldots$. Then, by (3.52), we find that

$$\|\beta_{j+1}\| = \left\| \left(\bar{Q}^{\dagger}\bar{Q}\mathscr{P}^n\right)^{\bar{k}} \beta_j \right\| \le \Lambda_2 \|\beta_j\| \quad \text{for all} \quad j = 0, 1, 2, \ldots. \tag{3.57}$$

From (3.54)–(3.57), we see that

$$\|\alpha_{j+1}\| \le \Lambda_1^{\bar{k}} \|\alpha_j\| + \left(\Lambda_2 + \left\|\Psi_{\varepsilon}\left(\bar{k}nT\right)\right\|\right) \|\beta_j\| \quad \text{for all} \quad j = 0, 1, 2, \ldots.$$

This, together with (3.57), implies that

$$\begin{pmatrix} \|\alpha_{j+1}\| \\ \|\beta_{j+1}\| \end{pmatrix} \le \begin{pmatrix} \Lambda_1^{\bar{k}} & \Lambda_2 + \left\|\Psi_{\varepsilon}\left(\bar{k}nT\right)\right\| \\ & \Lambda_2 \end{pmatrix} \begin{pmatrix} \|\alpha_j\| \\ \|\beta_j\| \end{pmatrix} \quad \text{for all} \quad j = 0, 1, 2, \ldots. \tag{3.58}$$

Because $\Lambda_1^{\bar{k}}$, $\Lambda_2 < 1$, it holds that $\lim_{j\to\infty} \alpha_j = \lim_{j\to\infty} \beta_j = 0$. Thus, we have that

$$\lim_{j\to\infty} y^{\varepsilon}\left(j\bar{k}nT; 0, x\right) = 0 \quad \text{for all} \quad x \in V^{\perp}, \tag{3.59}$$

since $y^{\varepsilon}\left(j\bar{k}nT; 0, x\right) = \alpha_j + \beta_j$. Finally, (3.46) follows from (3.51) and (3.59).

Step 3: To verify (c) $\Leftrightarrow \sigma\left(\bar{Q}^{\dagger}\bar{Q}\mathscr{P}^n\right) \subset \mathbb{B}$

Since V is an invariant subspace of $\mathscr{P}$ (see (3.10)), we have $V = V_1(\mathscr{P}) \oplus V_2(\mathscr{P})$ (see Remark 1.2). Let us recall (3.22). Because $V_1(\mathscr{P}) \subseteq \mathbb{R}_1^n(\mathscr{P})$ and $V_2(\mathscr{P}) \subseteq \mathbb{R}_2^n(\mathscr{P})$ (see Lemma 1.3), we can take $\{\xi_1, \ldots, \xi_{k_2}, \ldots, \xi_{k_3}\}$ as a basis of $\mathbb{R}_1^n(\mathscr{P})$, where $\Theta_1 \triangleq \{\xi_1, \xi_2, \ldots, \xi_{k_2}\}$ is a basis of $V_1(\mathscr{P})$; and take $\{\eta_1, \ldots, \eta_{k_1-k_2}, \ldots, \eta_{n-k_3}\}$ to be a basis of $\mathbb{R}_2^n(\mathscr{P})$, where $\Theta_3 \triangleq \{\eta_1, \eta_2, \ldots, \eta_{k_1-k_2}\}$ is a basis of $V_2(\mathscr{P})$. Write

$$\Theta_2 \triangleq \{\xi_{k_2+1}, \xi_{k_2+2}, \ldots, \xi_{k_3}\}, \qquad \Theta_4 \triangleq \{\eta_{k_1-k_2+1}, \eta_{k_1-k_2+2}, \ldots, \eta_{n-k_3}\}.$$

Since $V_1(\mathscr{P})$, $V_2(\mathscr{P})$, $\mathbb{R}_1^n(\mathscr{P})$ and $\mathbb{R}_2^n(\mathscr{P})$ are invariant subspaces of $\mathscr{P}$, there are matrices

$$A_1 \in \mathbb{R}^{k_2 \times k_2}, \quad A_{12} \in \mathbb{R}^{k_2 \times (k_3-k_2)}, \quad A_2 \in \mathbb{R}^{(k_3-k_2)\times(k_3-k_2)}$$

$$A_3 \in \mathbb{R}^{(k_1-k_2)\times(k_1-k_2)}, \quad A_{34} \in \mathbb{R}^{(k_1-k_2)\times(n-k_3-k_1+k_2)}, \quad A_4 \in \mathbb{R}^{(n-k_3-k_1+k_2)\times(n-k_3-k_1+k_2)},$$

so that

$$\mathscr{P}(\Theta_1, \Theta_2, \Theta_3, \Theta_4) = (\Theta_1, \Theta_2, \Theta_3, \Theta_4) \begin{pmatrix} A_1 & A_{12} & 0 & 0 \\ 0 & A_2 & 0 & 0 \\ 0 & 0 & A_3 & A_{34} \\ 0 & 0 & 0 & A_4 \end{pmatrix}. \tag{3.60}$$

Then, there are matrices $\hat{A}_{12} \in \mathbb{R}^{k_2 \times (k_3 - k_2)}$ and $\hat{A}_{34} \in \mathbb{R}^{(k_1-k_2)\times(n-k_3-k_1+k_2)}$ so that

$$\mathscr{P}^n(\Theta_1, \Theta_2, \Theta_3, \Theta_4) = (\Theta_1, \Theta_2, \Theta_3, \Theta_4) \begin{pmatrix} A_1^n & \hat{A}_{12} & 0 & 0 \\ 0 & A_2^n & 0 & 0 \\ 0 & 0 & A_3^n & \hat{A}_{34} \\ 0 & 0 & 0 & A_4^n \end{pmatrix}. \tag{3.61}$$

On the other hand, by (3.19) and (3.16), $I - \bar{Q}^\dagger \bar{Q}$ is a linear transform from $\mathbb{R}^n$ to $V = V_1(\mathscr{P}) \bigoplus V_2(\mathscr{P})$. Thus, there are vectors $c_1, \ldots, c_{k_1}$ in $\mathbb{R}^n$ so that

$$(I - \bar{Q}^\dagger \bar{Q})\zeta = \sum_{i=1}^{k_2} \langle \zeta, c_i \rangle \xi_i + \sum_{i=1}^{k_1-k_2} \langle \zeta, c_{k_2+i} \rangle \eta_i \quad \text{for all} \quad \zeta \in \mathbb{R}^n,$$

which leads to

$$\bar{Q}^\dagger \bar{Q}\zeta = \zeta - \sum_{i=1}^{k_2} \langle \zeta, c_i \rangle \xi_i - \sum_{i=1}^{k_1-k_2} \langle \zeta, c_{k_2+i} \rangle \eta_i \quad \text{for all} \quad \zeta \in \mathbb{R}^n.$$

Because $\bar{Q}V = \{0\}$ (see (3.16)), the above implies that there are matrices $C_1 \in \mathbb{R}^{k_2 \times (k_3-k_2)}$, $C_2 \in \mathbb{R}^{(k_1-k_2)\times(k_3-k_2)}$, $C_3 \in \mathbb{R}^{k_2 \times (n-k_3-k_1+k_2)}$, $C_4 \in \mathbb{R}^{(k_1-k_2)\times(n-k_3-k_1+k_2)}$ so that

$$\bar{Q}^\dagger \bar{Q}(\Theta_1, \Theta_2, \Theta_3, \Theta_4) = (\Theta_1, \Theta_2, \Theta_3, \Theta_4) \begin{pmatrix} 0 & C_1 & 0 & C_3 \\ 0 & I_{k_3-k_2} & 0 & 0 \\ 0 & C_2 & 0 & C_4 \\ 0 & 0 & 0 & I_{n-k_3-k_1+k_2} \end{pmatrix}. \tag{3.62}$$

From (3.60) and (3.62), it follows that

$$\bar{Q}^\dagger \bar{Q}\mathscr{P}(\Theta_1, \Theta_2, \Theta_3, \Theta_4) = (\Theta_1, \Theta_2, \Theta_3, \Theta_4) \begin{pmatrix} 0 & C_1 A_2 & 0 & C_3 A_4 \\ 0 & A_2 & 0 & 0 \\ 0 & C_2 A_2 & 0 & C_4 A_4 \\ 0 & 0 & 0 & A_4 \end{pmatrix},$$

which yields that

$$\sigma\left(\bar{Q}^{\dagger}\bar{Q}\mathscr{P}\right)=\begin{cases}\sigma(A_2)\bigcup\sigma(A_4)\bigcup\{0\}, & \text{if } k_1\geq 1;\\ \sigma(A_2)\bigcup\sigma(A_4), & \text{if } k_1=0.\end{cases} \tag{3.63}$$

Similarly, it follows from (3.61) and (3.62) that

$$\sigma\left(\bar{Q}^{\dagger}\bar{Q}\mathscr{P}^n\right)=\begin{cases}\sigma(A_2^n)\bigcup\sigma(A_4^n)\bigcup\{0\}, & \text{if } k_1\geq 1;\\ \sigma(A_2^n)\bigcup\sigma(A_4^n), & \text{if } k_1=0.\end{cases} \tag{3.64}$$

By (3.63) and (3.64), we see that

$$\sigma\left(\bar{Q}^{\dagger}\bar{Q}\mathscr{P}^n\right)=\left\{\lambda^n \mid \lambda\in\sigma\left(\bar{Q}^{\dagger}\bar{Q}\mathscr{P}\right)\right\}. \tag{3.65}$$

Therefore, we find that $\sigma\left(\bar{Q}^{\dagger}\bar{Q}\mathscr{P}\right)\subset\mathbb{B}\Leftrightarrow\sigma\left(\bar{Q}^{\dagger}\bar{Q}\mathscr{P}^n\right)\subset\mathbb{B}$. That is to say, $(c)\Leftrightarrow$ $\sigma\left(\bar{Q}^{\dagger}\bar{Q}\mathscr{P}^n\right)\subset\mathbb{B}$.

In summary, we have proved $(a)\Leftrightarrow(c)$ in Theorem 3.1.

3.2.2 *The Proof of (a)⇔(b) in Theorem* 3.1

It is clear that $(b)\Rightarrow(a)$. We will provide two methods to prove $(a)\Rightarrow(b)$. The first one is based on Theorem 1.4, while the second one is a direct construction method which provides an explicit feedback law. The first one is simpler than the second one, but does not give an explicit feedback law.

We begin with introducing the first method. Given a T-periodic pair $\left[A(\cdot),\,B(\cdot)\right]$, we denote by $(LQ)_{t,x}^{\infty}$, with $t\geq 0$ and $x\in\mathbb{R}^n$, the LQ problem with the system (3.2) and with the cost functional:

$$\int_t^{\infty}\left[\|u(s)\|^2+\|y(s;t,x,u)\|^2\right]ds,\quad u\in L^2(t,\infty;\mathbb{R}^m). \tag{3.66}$$

Notice that the above LQ problem is indeed the LQ problem (1.89) with $Q=I$ and $R=I$.

Lemma 3.5 *If a T-periodic pair $\left[A(\cdot),\,B(\cdot)\right]$ is nT-periodically stabilizable, then the above LQ problems $(LQ)_{0,x}^{\infty}$, satisfies the FCC for any $x\in\mathbb{R}^n$.*

Proof Let $\left[A(\cdot),\,B(\cdot)\right]$ be a T-periodic pair which is nT-periodically stabilizable. Then there exists an nT-periodic $K(\cdot)$ in $L^{\infty}(\mathbb{R}^+;\mathbb{R}^{m\times n})$ so that for some constants $C>0$ and $\delta>0$, $\|y_K(t;x)\|\leq Ce^{-\delta t}\|x\|$, when $t\geq 0$ and $x\in\mathbb{R}^n$, where $y_K(t;x)$ is the solution of (3.6) with the initial condition that $y(0)=x$. Given $x\in\mathbb{R}^n$, write $u^x(\cdot)=K(\cdot)y_K(\cdot;x)$. Then we have that $y(\cdot;0,x,u^x)=y_K(\cdot;x)$. Hence,

$$\int_t^{\infty}\left[\|u^x(s)\|^2+\|y(s;0,x,u^x)\|^2\right]ds \le \left(\|K\|^2_{L^\infty(\mathbb{R}^+;\mathbb{R}^{m\times n})}+1\right)\int_t^{\infty}\|y(s;0,x,u^x)\|^2 ds$$
$$\le \frac{C^2}{2\delta}\left(\|K\|^2_{L^\infty(\mathbb{R}^+;\mathbb{R}^{m\times n})}+1\right)<\infty.$$

Therefore, Problem $(LQ)^{\infty}_{0,x}$ satisfies the FCC for any $x\in\mathbb{R}^n$. This ends the proof. □

Proof of $(a)\Leftrightarrow(b)$ *in Theorem* 3.1 *(Method one).* By Lemma 3.5 and Theorem 1.4, we find that $(a)\Rightarrow(b)$. The reverse is obvious. This ends the proof. □

Proof of $(a)\Leftrightarrow(b)$ *in Theorem* 3.1 *(Method two).* Clearly, $(a)\Leftrightarrow(b)$ in the case that $n=1$. Thus, we can assume that $n\ge 2$. It is obvious that $(b)\Rightarrow(a)$. Now we show that $(a)\Rightarrow(b)$ for the case that $n\ge 2$. Suppose that (a) stands and $n\ge 2$. It suffices to show that there is a T-periodic stabilization law for $[A(\cdot),B(\cdot)]$. To verify this, we first construct a special $n\times n$ real matrix X (which appears in (3.26)), then find an $\varepsilon_0>0$ (depending on n, $\|X\|$ and $\|\mathscr{P}\|$), and finally prove that when $\varepsilon\in(0,\varepsilon_0]$, $K^{\varepsilon}(\cdot)$, given by (3.28) with the aforementioned X, is a T-periodic feedback stabilization law for $\left[A(\cdot),B(\cdot)\right]$. The detailed proof will be carried out by several steps.

Step 1: Structure of X in (3.26) *where $n\ge 2$*
Recall that $k_1=\dim V$ (see (3.22)). Hence, $\dim V^{\perp}=n-k_1$. We arbitrarily take a basis $\{\hat\eta_1,\dots,\hat\eta_{n-k_1}\}$ of $V^{\perp}$. The desired X will be given by

$$X\triangleq\left(\zeta_1,\dots,\zeta_{k_1},\hat\eta_1,\dots,\hat\eta_{n-k_1}\right). \tag{3.67}$$

Here, $\{\zeta_1,\dots,\zeta_{k_1}\}$ is a special basis of V, which will be determined later. Clearly, X is invertible. To construct the aforementioned basis $\{\zeta_1,\dots,\zeta_{k_1}\}$, we will build up subspaces $W_1, W_2,\dots$ and W_n of V so that

$$V_j=\bigoplus_{i=1}^{j}W_i \quad \text{for all } j\in\{1,2,\dots,n\} \tag{3.68}$$

and

$$\mathscr{P}W_{j+1}\subseteq W_j \quad \text{for all } \ j\in\{1,\dots n-1\}. \tag{3.69}$$

Here, we agree that $\{0\}$ is the 0-dimension subspace of $\mathbb{R}^n$.

When the above-mentioned $\{W_1,\dots,W_n\}$ is structured, it follows respectively from (3.68) and (3.69) that

$$V=V_n=\bigoplus_{i=1}^{n}W_i \tag{3.70}$$

and that $\dim W_{j+1}\le\dim W_j$ for each $j\in\{1,\dots n-1\}$. The latter implies that

$$W_{j+1}=\{0\},\quad \text{whenever } \ W_j=\{0\} \ \text{ for some } \ j\in\{1,\dots n-1\}. \tag{3.71}$$

Write $\{\hat{\zeta}_1, \dots \hat{\zeta}_{\hat{k}_1}\}$, $\{\hat{\zeta}_{\hat{k}_1+1}, \dots, \hat{\zeta}_{\hat{k}_2}\}$, ..., and $\{\hat{\zeta}_{\hat{k}_{n-1}+1}, \dots \hat{\zeta}_{\hat{k}_n}\}$ for bases of W_1, W_2, ..., and W_n, respectively. Here, we agree that any basis of W_j is $\emptyset$, if $W_j = \{0\}$. Since $k_1 = \dim V$, it follows from (3.70) and (3.71) that $\{\hat{\zeta}_1, \dots \hat{\zeta}_{k_1}\}$ is a basis of V. Then we take the desired basis $\{\zeta_1, \dots, \zeta_{k_1}\}$ in (3.67) to be $\{\hat{\zeta}_1, \dots \hat{\zeta}_{k_1}\}$.

The rest of this step is to structure $\{W_1, \dots, W_n\}$ satisfying (3.68) and (3.69). Two observations are given in order:

$$V_j = \mathscr{P}^{-1} V_{j-1} + V_{j-1} \quad \text{for all} \quad j \in \{2, \dots, n\} \tag{3.72}$$

and

$$V_j = V_{j-1} + \mathscr{P}^{-(j-1)} V_1 \quad \text{for all} \quad j \in \{2, \dots, n\}. \tag{3.73}$$

We now construct the above-mentioned $\{W_1, \dots, W_n\}$ by the following two cases:

In the first case when n = 2, we take $W_1 = V_1$. Then, from (3.73) where $j = 2$, we see that $V_2 = V_1 + \mathscr{P}^{-1} V_1$, by which, there is a subspace W_2 so that $V_2 = V_1 \bigoplus W_2$ and $W_2 \subset \mathscr{P}^{-1} V_1$. Hence, $\{W_1, W_2\}$ satisfies (3.68) and (3.69) in the case that $n = 2$.

In the second case that $n > 2$, we set $W_1 = V_1$. Then we build up $\{W_2, \dots W_n\}$ in such an order: $W_n \to W_{n-1} \to \cdots \to W_2$. By (3.73) where $j = n$, there is a subspace W_n so that

$$V_n = V_{n-1} \bigoplus W_n \quad \text{and} \quad W_n \subseteq \mathscr{P}^{-(n-1)} V_1. \tag{3.74}$$

Then, from the second property in (3.74), it follows that

$$\mathscr{P} W_n \subseteq \mathscr{P}^{-(n-2)} V_1. \tag{3.75}$$

Besides, it holds that

$$\mathscr{P} W_n \bigcap V_{n-2} = \{0\}. \tag{3.76}$$

In fact, if $y \in \mathscr{P} W_n \bigcap V_{n-2}$, then $y = \mathscr{P} z$ for some $z \in W_n$. Thus, $z = \mathscr{P}^{-1} y \in \mathscr{P}^{-1} V_{n-2}$. From this and (3.72), where $j = n - 1$, we find that $z \in V_{n-1}$. This, along with the facts that $z \in W_n$ and $V_{n-1} \bigcap W_n = \{0\}$ (see the first property in (3.74)), yields that $z = 0$. Hence, $y = \mathscr{P} z = 0$, which leads to (3.76).

We next build up W_{n-1}. From (3.73) where $j = n - 1$, (3.75) and (3.76), we see that

$$\begin{aligned} V_{n-1} &= V_{n-2} + \mathscr{P}^{-(n-2)} V_1 = V_{n-2} + \mathscr{P}^{-(n-2)} V_1 + \mathscr{P} W_n \\ &= \left(V_{n-2} \oplus \mathscr{P} W_n\right) + \mathscr{P}^{-(n-2)} V_1. \end{aligned}$$

Thus, there is a subspace $\hat{W}_{n-1}$ so that

$$\hat{W}_{n-1} \subseteq \mathscr{P}^{-(n-2)} V_1 \quad \text{and} \quad V_{n-1} = \left(V_{n-2} \bigoplus \mathscr{P} W_n\right) \bigoplus \hat{W}_{n-1}. \tag{3.77}$$

Let

$$W_{n-1} = \mathscr{P} W_n \bigoplus \hat{W}_{n-1}. \tag{3.78}$$

It is clear that

$$V_{n-1} = V_{n-2} \bigoplus W_{n-1} \text{ and } \mathscr{P} W_n \subseteq W_{n-1}. \tag{3.79}$$

Besides, from (3.78), (3.75) and the first result in (3.77), we obtain that

$$\mathscr{P} W_{n-1} \subseteq \mathscr{P}^{-(n-3)} V_1. \tag{3.80}$$

Meanwhile, since $V_{n-2} \bigcap W_{n-1} = \{0\}$ (see the first fact in (3.79)), from (3.72), with $j = n-2$, and using the same method to show (3.76), we can easily verify that

$$\mathscr{P} W_{n-1} \bigcap V_{n-3} = \{0\}. \tag{3.81}$$

By (3.80) and (3.81), following the same way to construct W_{n-1}, we can build up a subspace W_{n-2} with the similar properties as those in (3.79), (3.80) and (3.81). Then we can structure, step by step, subspaces $W_{n-3}, \ldots, W_2$ so that

$$V_j = V_{j-1} \bigoplus W_j \text{ and } \mathscr{P} W_{j+1} \subseteq W_j \text{ for all } j \in \{n-2, \ldots, 2\}. \tag{3.82}$$

Now, from the first property in (3.74), (3.79) and (3.82), noticing that $W_1 = V_1$, one can check that the aforementioned subspaces $W_1, \ldots, W_n$ satisfy (3.68) and (3.69) in the case that $n > 2$.

We end this step with the following property which will be used later:

$$\langle X^{-1} z_1, X^{-1} z_2 \rangle = 0 \ \forall \, z_1 \in W_i, \ z_2 \in W_j, \text{ with } i \neq j \text{ and } i, j \in \{1, 2, \ldots, n\}. \tag{3.83}$$

The property (3.83) can be easily verified, since $W_j \cap W_i = \{0\}$ for all $i \neq j$.

Step 2: Structure of a T-periodic $K^\varepsilon(\cdot)$ in $L^\infty(\mathbb{R}^+; \mathbb{R}^{m\times n})$ and a positive number ε_0

Let X be given by (3.67). For each $\varepsilon > 0$, $t \in [0, T)$ and $x \in \mathbb{R}^n$, we consider the LQ Problem $(LQ)_{t,x}^{T,\varepsilon}$, with the system (3.2) over $[0, T]$, and with the cost functional:

$$J_{t,x}^{T,\varepsilon}(u) = \int_t^T \varepsilon \|u(s)\|^2 \mathrm{d}t + \|X^{-1} \mathscr{P}^{-1} y(T; t, x, u)\|^2. \tag{3.84}$$

According to Theorem 1.2, the value function $W^{T,\varepsilon}$ associated with the problem $\{(LQ)_{t,x}^{T,\varepsilon}\}_{(t,x)\in[0,T)\times\mathbb{R}^n}$ satisfies that

$$W^{T,\varepsilon}(t, x) = \langle x, \hat{\Upsilon}^\varepsilon(t) x \rangle \text{ for all } (t, x) \in [0, T) \times \mathbb{R}^n, \tag{3.85}$$

where $\hat{\Upsilon}^{\varepsilon}(\cdot)$ is the solution to the following Riccati equation:

$$\begin{cases} \dot{\Upsilon}(t) + A(t)^*\Upsilon(t) + \Upsilon(t)A(t) - \frac{1}{\varepsilon}\Upsilon(t)B(t)B(t)^*\Upsilon(t)^* = 0, & t \in [0, T]; \\ \Upsilon(T) = \mathscr{P}^{-*}X^{-*}X^{-1}\mathscr{P}^{-1}. \end{cases}$$

One can finds that $S^{\varepsilon}(\cdot) = \left(\hat{\Upsilon}^{\varepsilon}(\cdot)\right)^{-1}$, where $S^{\varepsilon}(\cdot)$ is the unique solution to the Eq. (3.26).

By (3.8), for each $x \in V_1$, there is a control $u_1 \in \mathscr{U}_{ad}$ so that $y(T; 0, x, u_1) = 0$. Thus, by (3.85), we find that

$$0 \leq \lim_{\varepsilon \to 0^+} \langle x, \hat{\Upsilon}^{\varepsilon} x \rangle = \lim_{\varepsilon \to 0^+} W^{T,\varepsilon}(0, x) \leq \lim_{\varepsilon \to 0^+} J_{0,x}^{T,\varepsilon}(u_1) = 0, \quad \text{when } x \in V_1.$$

From this, $\lim_{\varepsilon \to 0^+} \langle x, \hat{\Upsilon}^{\varepsilon}(0)x \rangle = 0$ for each $x \in V_1$. Since V_1 is of finite dimension,

$$\langle x, \hat{\Upsilon}^{\varepsilon}(0)x \rangle \to 0 \ \text{ uniformly in } \left\{x \in V_1 : \|x\| \leq 1\right\}. \tag{3.86}$$

Let

$$\delta \triangleq \frac{1}{2n\|X\|\left(1 + \left(\sqrt{2}\|\mathscr{P}\|\|X\|\|X^{-1}\|\right)^n\right)}. \tag{3.87}$$

By (3.86), there is an

$$\varepsilon_0 \triangleq \varepsilon_0(\delta) \triangleq \varepsilon_0(n, X, \mathscr{P}) > 0 \tag{3.88}$$

so that when $\varepsilon \in (0, \varepsilon_0]$,

$$\langle x, \hat{\Upsilon}^{\varepsilon}(0)x \rangle \leq \delta^2\|x\|^2 \ \text{ for all } \ x \in V_1. \tag{3.89}$$

In the rest of the proof, we fix an $\varepsilon \in (0, \varepsilon_0]$. Let $K^{\varepsilon}(\cdot)$ be defined by (3.28), where X is given by (3.67). Write $\bar{y}^x(\cdot)$ for the solution to the equation:

$$\begin{cases} \dot{y}(t) = \left(A(t) + B(t)K^{\varepsilon}(t)\right)y(t), & \text{a.e. } t \in \mathbb{R}^+, \\ y(0) = x. \end{cases} \tag{3.90}$$

By Theorem 1.2, the control

$$\bar{u}^x(\cdot) \triangleq K^{\varepsilon}(\cdot)\bar{y}^x(\cdot), \tag{3.91}$$

when it is restricted over $(0, T)$, is the optimal control to Problem $(LQ)_{0,x}^{T,\varepsilon}$.

Step 3: To prove that the above $K^{\varepsilon}(\cdot)$ is a T-periodic feedback stabilization law Define a linear mapping $\mathscr{L}$ on $\mathbb{R}^n$ by

$$\mathscr{L}(x) = x + \int_0^T \widehat{\Phi}^{-1}(s)B(s)\bar{u}^x(s)\mathrm{d}s \quad \text{for all} \quad x \in \mathbb{R}^n, \tag{3.92}$$

where $\bar{u}^x(\cdot)$ is given by (3.91). Clearly,

$$\bar{y}^x(T) = \mathscr{P}\mathscr{L}(x) \quad \text{for all} \quad x \in \mathbb{R}^n. \tag{3.93}$$

First, we claim that

$$\mathscr{L}(x) = x, \quad \text{if } x \in \bigoplus_{j=2}^{n} W_j; \qquad \mathscr{L}(x) \in W_1 \ \text{ and } \ \|\mathscr{L}(x)\| \le \|x\|, \quad \text{if } x \in W_1. \tag{3.94}$$

To prove the first statement in (3.94), it suffices to show that $\bar{u}^x(\cdot) \equiv 0$ for all $x \in \bigoplus_{j=2}^n W_j$. For this purpose, we observe from (3.11) and (3.68) that

$$z \triangleq \int_0^T \widehat{\Phi}^{-1}(s)B(s)u(s)\mathrm{d}s \in V_1 = W_1 \quad \text{for al} \quad u \in L^2(0, T; \mathbb{R}^m).$$

We also find that $y(T; 0, x, u) = \mathscr{P}(x + z)$ for all $x \in \mathbb{R}^n$. These, together with (3.83), yields that when $x \in \bigoplus_{j=2}^n W_j$,

$$\begin{aligned} J_{0,x}^{T,\varepsilon}(u) &\ge \langle X^{-1}\mathscr{P}^{-1}y(T; 0, x, u), X^{-1}\mathscr{P}^{-1}y(T; 0, x, u)\rangle \\ &= \langle X^{-1}(x + z), X^{-1}(x + z)\rangle \ge \langle X^{-1}x, X^{-1}x\rangle = J_{0,x}^{T,\varepsilon}(0). \end{aligned}$$

Hence, the null control is the optimal control to Problem $(LQ)_{0,x}^{T,\varepsilon}$. Since the optimal control to this problem is unique (see Lemma 1.6), it stands that $\bar{u}^x(\cdot) \equiv 0$.

To prove the second statement in (3.94), we observe from (3.11) that $\mathscr{L}W_1 \subseteq W_1$ (since $W_1 = V_1$). Then, by the optimality of $\bar{u}^x$, we see that

$$J_{0,x}^{T,\varepsilon}(\bar{u}^x) = \langle x, \hat{\Upsilon}^\varepsilon(0)x\rangle \quad \text{for all} \quad x \in \mathbb{R}^n.$$

This, together with the definition of $J_{0,x}^{T,\varepsilon}$ and (3.89), indicates that

$$\|X^{-1}\mathscr{P}^{-1}\bar{y}^x(T)\| \le \sqrt{\hat{J}^\varepsilon(\bar{u}^x, x)} = \sqrt{\langle x, \hat{\Upsilon}^\varepsilon(0)x\rangle} \le \delta\|x\| \quad \text{for each} \quad x \in W_1 = V_1.$$

From this and (3.93), we see that

$$\|\mathscr{L}(x)\| = \|\mathscr{P}^{-1}\bar{y}^x(T)\| \le \|X\|\left\|X^{-1}\mathscr{P}^{-1}\bar{y}^x(T)\right\| \le \delta\|X\|\|x\| \quad \text{for each} \quad x \in W_1, \tag{3.95}$$

which, along with (3.87), leads to the second statement in (3.94).

Now we conclude that the claim (3.94) stands. In addition, by (3.95), we see that

$$\|\bar{y}^x(T)\| = \|\mathscr{P}\mathscr{L}(x)\| \le \|\mathscr{P}\|\|\mathscr{L}(x)\| \le \delta\|\mathscr{P}\|\|X\|\|x\| \quad \text{for each} \quad x \in W_1. \tag{3.96}$$

Second, we claim that

$$\|\bar{y}^x(T)\| \leq \sqrt{2}\|\mathscr{P}\|\|X\|\|X^{-1}\|\|x\| \quad \text{for each} \quad x \in V. \tag{3.97}$$

In fact, by (3.70), each $x \in V$ can be expressed as: $x = x_1 + x_2$, where $x_1 \in W_1$ and $x_2 \in \bigoplus_{j=2}^{n} W_j$. Because of (3.83), vectors $X^{-1}x_1$ and $X^{-1}x_2$ are orthogonal. Thus,

$$\|X^{-1}x_1 + X^{-1}x_2\|^2 = \|X^{-1}x_1\|^2 + \|X^{-1}x_2\|^2 \geq 1/2\big(\|X^{-1}x_1\| + \|X^{-1}x_2\|\big)^2.$$

This, along with the second statement in (3.94), yields that

$$\begin{aligned}\|\bar{y}^x(T)\| &= \|\bar{y}^{x_1}(T) + \bar{y}^{x_2}(T)\| = \|\mathscr{P}\big(\mathscr{L}(x_1) + \mathscr{L}(x_2)\big)\| \\ &\leq \sqrt{2}\|\mathscr{P}\|\|X\|\big(\|X^{-1}x_1 + X^{-1}x_2\|\big),\end{aligned}$$

which leads to (3.97).

Next, we claim that

$$\|y^x(nT)\| \leq \delta\|\mathscr{P}\|^n(\sqrt{2}\|X\|\|X^{-1}\|)^{n-1}\|X\|\|x\| \quad \text{for all} \quad x \in \bigcup_{i=1}^{n} W_i. \tag{3.98}$$

For this purpose, we first observe from the T-periodicity of $A(\cdot) + B(\cdot)K^{\varepsilon}(\cdot)$ that

$$\bar{y}^x(kT) = \bar{y}^{\bar{y}^x((k-1)T)}(T) \quad \text{for all} \quad x \in \mathbb{R}^n \quad \text{and} \quad k \in \mathbb{N}. \tag{3.99}$$

From (3.99) and (3.97), we find that

$$\|\bar{y}^x(nT)\| \leq \Big(\sqrt{2}\|\mathscr{P}\|\|X\|\|X^{-1}\|\Big)^{n-1}\|\bar{y}^x(T)\| \quad \text{for all} \quad x \in V.$$

This, along with (3.96), indicates that the estimate in (3.98) holds when $x \in W_1$. We now prove (3.98) for the case when $x \in W_j$ for some $j \in \{2, 3, \ldots, n\}$. By (3.99), (3.93), (3.94) and (3.69), using the mathematical induction, one can easily check that

$$\bar{y}^x((j-1)T) = \mathscr{P}\bar{y}^x((j-2)T) = \mathscr{P}^{j-1}x \in W_1.$$

This, together with (3.99) and (3.96), yields that

$$\|\bar{y}^x(jT)\| \leq \delta\|\mathscr{P}\|\|X\|\|\bar{y}^x((j-1)T)\| \leq \delta\|\mathscr{P}\|^j\|X\|\|x\| \text{ for all } x \in W_j. \tag{3.100}$$

On the other hand, from (3.97), it follows that

$$\|\bar{y}^z(kT)\| \leq \Big(\sqrt{2}\|\mathscr{P}\|\|X\|\|X^{-1}\|\Big)^k\|z\| \quad \text{for all} \quad z \in V \quad \text{and} \quad k \in \mathbb{N}.$$

Since $\bar{y}^x(nT) = \bar{y}^{\bar{y}^x(jT)}((n-j)T)$, the above inequality yields

$$\|\bar{y}^x(nT)\| \le \left(\sqrt{2}\|\mathscr{P}\|\|X\|\|X^{-1}\|\right)^{n-j} \|\bar{y}^x(jT)\|,$$

which, together with (3.100), shows that

$$\begin{aligned}\|\bar{y}^x(nT)\| &\le \delta\|\mathscr{P}\|^n\left(\sqrt{2}\|X\|\|X^{-1}\|\right)^{n-j}\|X\|\|x\| \\ &< \delta\|\mathscr{P}\|^n\left(\sqrt{2}\|X\|\|X^{-1}\|\right)^{n-1}\|X\|\|x\|.\end{aligned}$$

Since x was arbitrarily taken from one of W_j with $j \in \{2, 3, \ldots, n\}$, the above estimate holds for all $x \in \bigcup_{i=2}^{n} W_i$. Thus, we have proved (3.98).

Then, we claim that

$$\|\bar{y}^x(nT)\| \le \delta\|x\| \quad \text{for all } x \in V. \tag{3.101}$$

In fact, by (3.70), each $x \in V$ can be expressed as: $x = \sum_{j=1}^{n} x_j$ with $x_j \in W_j$ for all $j = 1, \ldots, n$. Thus, we find that

$$\|\bar{y}^x(nT)\| = \left\|\sum_{j=1}^{n} \bar{y}^{x_j}(nT)\right\| \le \sum_{j=1}^{n} \|\bar{y}^{x_j}(nT)\|.$$

This, as well as (3.98), yields that

$$\|\bar{y}^x(nT)\| \le \delta\|\mathscr{P}\|^n(\sqrt{2}\|X\|\|X^{-1}\|)^{n-1}\|X\|\sum_{j=1}^{n}\|x_j\|.$$

Meanwhile, since $\langle X^{-1}x_i, X^{-1}x_j\rangle = 0$ when $i \ne j$ (see (3.83)), one finds that

$$n\left\|\sum_{j=1}^{n} X^{-1}x_j\right\|^2 \ge \left(\sum_{j=1}^{n}\|X^{-1}x_j\|\right)^2.$$

Hence,

$$\|\bar{y}^x(nT)\| < \delta n\|X\|(\sqrt{2}\|\mathscr{P}\|\|X\|\|X^{-1}\|)^n\|x\|.$$

This, together with (3.87), leads to (3.101).

Write $\hat{\mathscr{P}}_\varepsilon$ for the periodic map associated with $A(\cdot) + B(\cdot)K^\varepsilon(\cdot)$. It is clear that

$$\bar{y}^x(nT) = \left(\hat{\mathscr{P}}_\varepsilon\right)^n x \quad \text{for each } x \in \mathbb{R}^n, \tag{3.102}$$

and

$$\bar{y}^x(T) = \mathscr{P}\Big[x + \int_0^T \widehat{\Phi}^{-1}(s)B(s)\bar{u}^x(s)\mathrm{d}s\Big] \text{ for each } x \in \mathbb{R}^n. \tag{3.103}$$

Here, we recall that $\widehat{\Phi}(\cdot)$ is the fundamental solution for $A(\cdot)$ and $\bar{u}_\varepsilon^x(\cdot)$ is given by (3.91). Then, by (3.11), we see that

$$x + \int_0^T \widehat{\Phi}^{-1}(s)B(s)\bar{u}^x(s)\mathrm{d}s \in V, \quad \text{when } x \in V.$$

This, along with (3.103) and (3.10), yields that $\bar{y}^x(T) \in V$, when $x \in V$. Using (3.10) again, we see that $\bar{y}^x(2T) = \bar{y}^{\bar{y}^x(T)}(T) \in V$. Then, step by step, we reach that

$$\bar{y}^x(nT) = \bar{y}^{\bar{y}^x((n-1)T)}(T) \in V, \quad \text{when } x \in V.$$

This leads to

$$(\hat{\mathscr{P}}_\varepsilon)^n : V \to V. \tag{3.104}$$

Hence, there are $\tilde{A}_{11} \in \mathbb{R}^{k_1\times k_1}$, $\tilde{A}_{21} \in \mathbb{R}^{k_1\times(n-k_1)}$, $\tilde{A}_{22} \in \mathbb{R}^{(n-k_1)\times(n-k_1)}$ so that

$$\big(\hat{\mathscr{P}}_\varepsilon\big)^n X = X \begin{pmatrix} \tilde{A}_{11} & \tilde{A}_{12} \\ 0 & \tilde{A}_{22} \end{pmatrix}. \tag{3.105}$$

By (3.105), we find that $\sigma((\hat{\mathscr{P}}_\varepsilon)^n|_V) = \sigma(\tilde{A}_{11})$. Then, from (3.102) and (3.101), we see that if $\delta < 1$ is given by (3.87), then $\|(\hat{\mathscr{P}}_\varepsilon)^n x\| \le \delta\|x\|$ for all $x \in V$. Since V is invariant under $\big(\hat{\mathscr{P}}_\varepsilon\big)^n$ (see (3.104)), the above inequality implies that $\|(\hat{\mathscr{P}}_\varepsilon)^n|_V\| < 1$. Furthermore, it follows that $\sigma((\hat{\mathscr{P}}_\varepsilon)^n|_V) \subset \mathbb{B}$ (see Appendix C5 in [86]). Hence, $\sigma(\tilde{A}_{11}) \subset \mathbb{B}$.

We next prove that $\sigma(\tilde{A}_{22}) \subset \mathbb{B}$. When this is done, it follows from (3.105) that $\sigma((\hat{\mathscr{P}}_\varepsilon)^n) \subset \mathbb{B}$, which leads to the facts that $\sigma(\hat{\mathscr{P}}_\varepsilon) \subset \mathbb{B}$ and K^ε is a T-periodic feedback stabilization law for the pair $[A(\cdot), B(\cdot)]$. Thus, we complete the proof of the statement that $(a) \Leftrightarrow (b)$.

The proof of $\sigma(\tilde{A}_{22}) \subset \mathbb{B}$ is as follows. Recall that k_1 and $X = \big(\zeta_1, \ldots, \zeta_{k_1}, \hat{\eta}_1, \ldots, \hat{\eta}_{n-k_1}\big)$ are given by (3.22) and (3.67), respectively. Since $\{\hat{\eta}_1, \ldots, \hat{\eta}_{n-k_1}\}$ is a basis of $V^\perp$ (see (3.67)), the matrix $(\hat{\eta}_1, \ldots, \hat{\eta}_{n-k_1})$ can be treated as a linear and one-to-one map from $\mathbb{R}^{n-k_1}$ to $V^\perp$. Write

$$z(c) \triangleq (\hat{\eta}_1, \ldots, \hat{\eta}_{n-k_1})c \in V^\perp \text{ for each } c \in \mathbb{R}^{n-k_1}. \tag{3.106}$$

By (3.106) and (3.67), we see that

$$z(c) = X\begin{pmatrix} 0_{k_1\times k_1} \\ c \end{pmatrix} \text{ for all } c \in \mathbb{R}^{n-k_1}. \tag{3.107}$$

From (3.102), (3.107) and (3.105), it follows that

$$\bar{y}^{z(c)}(nT) = \left(\hat{\mathscr{P}}_\varepsilon\right)^n z(c) = (\zeta_1, \ldots, \zeta_{k_1})\tilde{A}_{12}c + (\hat{\eta}_1, \ldots, \hat{\eta}_{n-k_1})\tilde{A}_{22}c \quad \text{for all } c \in \mathbb{R}^{n-k_1}. \tag{3.108}$$

Since $\{\zeta_1, \ldots, \zeta_{k_1}\}$ and $\{\hat{\eta}_1, \ldots, \hat{\eta}_{n-k_1}\}$ are respectively the bases of V and $V^\perp$, we have

$$(\zeta_1, \ldots, \zeta_{k_1})\tilde{A}_{12}c \in V \quad \text{and} \quad (\hat{\eta}_1, \ldots, \hat{\eta}_{n-k_1})\tilde{A}_{22}c \in V^\perp \quad \text{for all } c \in \mathbb{R}^{n-k_1}. \tag{3.109}$$

Let $\left(\bar{y}_1^{z(c)}(\cdot), \bar{y}_2^{z(c)}(\cdot)\right)$ be the unique decomposition of $\bar{y}^{z(c)}(\cdot)$ provided by Lemma 3.4, where $y(\cdot)$ and $u(\cdot)$ are replaced by $\bar{y}^{z(c)}(\cdot)$ and $\bar{u}^{z(c)}(\cdot)$ respectively. From the periodicity of (3.90), and from (3.18) and (3.10), we see that

$$\bar{y}_1^{z(c)}(nT) \in V \quad \text{and} \quad \bar{y}_2^{z(c)}(nT) \in V^\perp \quad \text{for all } c \in \mathbb{R}^{n-k_1}.$$

These, together with (3.108) and (3.109), indicate that

$$\bar{y}_2^{z(c)}(nT) = (\hat{\eta}_1, \ldots, \hat{\eta}_{n-k_1})\tilde{A}_{22}c \quad \text{for all } c \in \mathbb{R}^{n-k_1}. \tag{3.110}$$

Meanwhile, it follows from (3.20) and (3.106) that for all $c \in \mathbb{R}^{n-k_1}$,

$$\bar{y}_2^{z(c)}(nT) = \bar{Q}^\dagger \bar{Q}\mathscr{P}^n \bar{y}_2^{z(c)}(0) = \bar{Q}^\dagger \bar{Q}\mathscr{P}^n z(c) = \bar{Q}^\dagger \bar{Q}\mathscr{P}^n (\hat{\eta}_1, \ldots, \hat{\eta}_{n-k_1})c.$$

From (3.110) and the above, we see that

$$(\hat{\eta}_1, \ldots, \hat{\eta}_{n-k_1})\tilde{A}_{22} = \bar{Q}^\dagger \bar{Q}\mathscr{P}^n (\hat{\eta}_1, \ldots, \hat{\eta}_{n-k_1}). \tag{3.111}$$

On the other hand, because $V^\perp$ is invariant under $\bar{Q}^\dagger \bar{Q}\mathscr{P}^n$ (see (3.20)), there is a matrix $\tilde{A}_2 \in \mathbb{R}^{(n-k_1)\times(n-k_1)}$ so that

$$\bar{Q}^\dagger \bar{Q}\mathscr{P}^n (\hat{\eta}_1, \ldots, \hat{\eta}_{n-k_1}) = (\hat{\eta}_1, \ldots, \hat{\eta}_{n-k_1})\tilde{A}_2 \quad \text{and} \quad \sigma(\tilde{A}_2) \subseteq \sigma\left(\bar{Q}^\dagger \bar{Q}\mathscr{P}^n\right). \tag{3.112}$$

By (3.111) and the first equality in (3.112), we find that $\tilde{A}_2 = \tilde{A}_{22}$. Since we already proved $(a) \Leftrightarrow \sigma\left(\bar{Q}^\dagger \bar{Q}\mathscr{P}^n\right) \subset \mathbb{B}$ in Sect. 3.2.1, and because we are in the case that (a) is assumed to be true, it follows from the second inclusion in (3.112) that $\sigma(\tilde{A}_2) \subset \mathbb{B}$. Thus, $\sigma(\tilde{A}_{22}) \subset \mathbb{B}$.

In summary, we have finished the proof of Theorem 3.1. □

Remark 3.3 The periodic stabilization criterion (c) in Theorem 3.1 is an extension of Kalman's stabilization criterion for time invariant pairs in $\mathbb{R}^{n\times n} \times \mathbb{R}^{n\times m}$. To see it, we let $[A, B] \in \mathbb{R}^{n\times n} \times \mathbb{R}^{n\times m}$, with the null-controllable subspace V. Write $\{\xi_1, \xi_2, \ldots, \xi_{k_1}\}$ and $\{\xi_{k_1+1}, \xi_{k_1+2}, \ldots, \xi_n\}$ for normalized orthogonal bases of V and

$V^\perp$ respectively. Let $\hat{Q} = (\xi_1, \ldots, \xi_{k_1}, \ldots, \xi_n)$. By the classical linear control theory (see, for instance, Theorem 1.6, p. 110, [59]), there are matrices $A_1 \in \mathbb{R}^{k_1 \times k_1}$, $A_2 \in \mathbb{R}^{k_1 \times (n-k_1)}$, $A_3 \in \mathbb{R}^{(n-k_1)\times(n-k_1)}$ and $B_1 \in \mathbb{R}^{k_1 \times m}$, with $[A_1, B_1]$ controllable, so that

$$A = \hat{Q} \begin{pmatrix} A_1 & A_2 \\ 0 & A_3 \end{pmatrix} \hat{Q}^*, \qquad B = \hat{Q} \begin{pmatrix} B_1 \\ 0 \end{pmatrix}.$$

Furthermore, $[A, B]$ is stabilizable if and only if A_3 is exponentially stable; while A_3 is exponentially stable if and only if $\big[A, B\big]$ satisfies the Kalman's stabilization condition (see, for instance, [59] or [86]).

Let $\widehat{\Phi}(\cdot)$ be the fundamental solution associated with A, and $\bar{Q}$ be the matrix defined in Theorem 3.1 where $\big[A(\cdot), B(\cdot)\big]$ is replaced by $\big[A, B\big]$. By a direct calculation, we have that for any $\hat{T} > 0$,

$$\widehat{\Phi}(\hat{T}) = \hat{Q} \begin{pmatrix} e^{\hat{T}A_1}, & \int_0^{\hat{T}} e^{(\hat{T}-s)A_1} A_2 e^{sA_3} ds \\ 0 & e^{\hat{T}A_3} \end{pmatrix} \hat{Q}^*$$

and

$$\bar{Q} = \hat{Q} \begin{pmatrix} 0 & 0 \\ 0 & e^{\hat{T}A_3^*} e^{\hat{T}A_3} \end{pmatrix} \hat{Q}^*.$$

Thus, it holds that

$$\bar{Q}^\dagger \bar{Q} \widehat{\Phi}(\hat{T}) = \hat{Q} \begin{pmatrix} 0 & 0 \\ 0 & e^{\hat{T}A_3} \end{pmatrix} \hat{Q}^*.$$

Hence, when $\big[A(\cdot), B(\cdot)\big] \equiv [A, B]$ is time invariant, we have that for any $\hat{T} > 0$

$$\begin{aligned} \sigma(\bar{Q}^\dagger \bar{Q} \widehat{\Phi}(\hat{T})) \subset \mathbb{B} \Leftrightarrow \sigma(e^{\hat{T}A_3}) \subset \mathbb{B} \Leftrightarrow \sigma(A_3) \subset \mathbb{C}^- \\ \Leftrightarrow \big[A, B\big] \text{ holds Kalman's condition.} \end{aligned}$$

From these, we see that if $\big[A(\cdot), B(\cdot)\big] = [A, B]$ is time invariant, then (3.1) is $\hat{T}$-periodically stabilizable for some $\hat{T} > 0$ if and only if (3.1) is $\hat{T}$-periodically stabilizable for any $\hat{T} > 0$ if and only if (3.1) is feedback stabilizable by a constant matrix.

3.2.3 Decay Rate of Stabilized Equations

Based on Theorem 3.1, we can get an estimate on the decay rate for solutions of Eq. (3.6) where $[A(\cdot), B(\cdot)]$ is T-periodically stabilizable and $K(\cdot) = K_n^\varepsilon(\cdot)$ with ε sufficiently small. To state it, we let

$$\rho_0 \triangleq \min_{\lambda\in\sigma(\bar{Q}^\dagger\bar{Q}\mathscr{P})\setminus\{0\}} -\ln|\lambda|/T. \tag{3.113}$$

Theorem 3.2 *Let $[A(\cdot), B(\cdot)]$ be a T-periodically stabilizable pair with V given by (3.7). Then, given $\delta > 0$, there are positive numbers $\varepsilon \triangleq \varepsilon(\delta)$ and $C \triangleq C(\delta)$ so that any solution $y^\varepsilon(\cdot)$ to (3.6) with $K(\cdot) = K_n^\varepsilon(\cdot)$ (given by (3.27)) satisfies that*

$$\|y^\varepsilon(s)\| \le M\left(e^{-s/\delta}\left\|\mathrm{Proj}_V\left(y^\varepsilon(0)\right)\right\| + e^{-(\bar{\rho}-\delta)s}\left\|\mathrm{Proj}_{V^\perp}\left(y^\varepsilon(0)\right)\right\|\right), \quad s \ge 0. \tag{3.114}$$

Proof We organize the proof by the following two cases: In the first case where $V = \mathbb{R}^n$, it follows from the second equality in (3.16) that $\bar{Q} = 0_{n\times n}$. This, together with (3.113), yields that $\rho_0 = +\infty$. Thus, the estimate (3.114) is equivalent to the estimate:

$$\|y^\varepsilon(s)\| \le Me^{-s/\delta}\|y^\varepsilon(0)\|, \quad s \ge 0, \tag{3.115}$$

for any solution $y^\varepsilon(\cdot)$ to Eq. (3.6) with $K(\cdot) = K_n^\varepsilon(\cdot)$. By (3.15) and (3.44), we find

$$\lim_{\varepsilon\to 0+} \Lambda_1(\varepsilon) \triangleq \lim_{\varepsilon\to 0+} \|(S_n^\varepsilon(0))^{-1} - \bar{Q}\|^{\frac{1}{2}} = 0.$$

Then, given $\delta > 0$, there is an $\varepsilon \triangleq \varepsilon(\delta) > 0$ so that $\Lambda_1(\varepsilon) \le e^{-nT/\delta}$. Notice that any solution $y^\varepsilon(\cdot)$ to Eq. (3.6) with $K(\cdot) = K_n^\varepsilon(\cdot)$ satisfies that $y^\varepsilon(0) \in V$ in this case. Then, by (3.50), where $\Lambda_1 = \Lambda_1(\varepsilon)$ (see (3.44)), we find that

$$\|y^\varepsilon(jnT)\| \le e^{-jnT/\delta}\|y^\varepsilon(0)\| \ \text{ for all } \ y^\varepsilon(0) \in V \ \text{ and } \ j \in \mathbb{N}. \tag{3.116}$$

Write $\widehat{\Phi}^\varepsilon(\cdot)$ for the fundamental solution associated with $A(\cdot) + B(\cdot)K_n^\varepsilon(\cdot)$. Let

$$M_1^\varepsilon \triangleq \sup_{r\in[0,nT]} \left\|\widehat{\Phi}^\varepsilon(r)\left(\widehat{\Phi}^\varepsilon(nT)\right)^{-1}\right\|. \tag{3.117}$$

By (3.117) and (3.116), we see that

$$\begin{aligned}\|y^\varepsilon(s)\| &= \left\|\widehat{\Phi}^\varepsilon(s)\left(\widehat{\Phi}^\varepsilon\left(\left(\left[\tfrac{s}{nT}\right]+1\right)nT\right)^{-1}\widehat{\Phi}^\varepsilon\left(\left(\left[\tfrac{s}{nT}\right]+1\right)nT\right)y^\varepsilon(0)\right\| \\ &\le M_1^\varepsilon e^{-(nT/\delta)\left(\left[\frac{s}{nT}\right]+1\right)}\|y^\varepsilon(0)\| \le M_1^\varepsilon e^{-s/\delta}\|y^\varepsilon(0)\| \ \text{ for each } \ s > 0.\end{aligned} \tag{3.118}$$

Since (3.115)$\Leftrightarrow$ (3.114) in this case, (3.118) leads to (3.114) in the case that $V = \mathbb{R}^n$.

In the second case where $V \ne \mathbb{R}^n$, we find from the second equality in (3.16) that $\bar{Q} \ne 0_{n\times n}$. This implies that $\bar{Q}^\dagger\bar{Q}\mathscr{P} \ne 0_{n\times n}$. So $\bar{Q}^\dagger\bar{Q}\mathscr{P}$ has a nonzero eigenvalue. Thus, $\rho_0 < +\infty$. Meanwhile, since $[A(\cdot), B(\cdot)]$ is T-periodically stabilizable, it follows from Theorem 3.1 that $\sigma\left(\bar{Q}^\dagger\bar{Q}\mathscr{P}\right) \subset \mathbb{B}$. Hence, $\rho_0 > 0$ (see (3.113)).

Given $\delta > 0$, we take

$$\rho_2 = \max\{\rho_0 - \delta,\ \rho_0/2\} \ \text{ and } \ \hat{\lambda} = e^{-\rho_2 T}. \tag{3.119}$$

Since $0 < \rho_2 < \rho_0$, it follows from (3.113) and the second property in (3.119) that

$$\max\left\{|\lambda| \mid \lambda \in \sigma\left(\bar{Q}^\dagger \bar{Q}\mathscr{P}\right)\right\} < \hat{\lambda} < 1.$$

This, together with (3.65), implies that the spectral radius of $\bar{Q}^\dagger \bar{Q}\mathscr{P}^n$ is less than 1. By the equivalent definition of spectral radius (see [97]), there is a $\bar{k} \in \mathbb{N}$ so that

$$\Lambda_2 \triangleq \left\|\left(\bar{Q}^\dagger \bar{Q}^n \mathscr{P}\right)^{\bar{k}}\right\| \le \hat{\lambda}^{\bar{k}n} = e^{-\rho_2 \bar{k}nT} < 1. \tag{3.120}$$

We now set

$$\rho_1 \triangleq \max\{\rho_2 + 1,\ 1/\delta\} \text{ and } \Lambda_1 \triangleq e^{-\rho_1 nT}. \tag{3.121}$$

By the same argument as that used to prove (3.116), we can find an $\varepsilon \triangleq \varepsilon(\delta) > 0$ so that $\Lambda_1(\varepsilon) \le \Lambda_1$ and so that $\|y^\varepsilon(jnT)\| \le e^{-j\rho_1 nt}\|y^\varepsilon(0)\| = \Lambda_1^j\|y^\varepsilon(0)\|$ for all $j \in \mathbb{N}$, when $y^\varepsilon(0) \in V$. Then, by the above equation with $j = 1$, we see that

$$\|y^\varepsilon(nT)\| \le e^{-\rho_1 nT}\|y^\varepsilon(0)\| = \Lambda_1\|y^\varepsilon(0)\| \text{ for all } y^\varepsilon(0) \in V. \tag{3.122}$$

Notice that (3.122) and (3.120) correspond to respectively (3.47) (with different Λ_1) and (3.52) (with different Λ_2 and $\bar{k}$).

Set $y^\varepsilon(0) = x$. Write $(y_1(\cdot; 0, x), y_2(\cdot; 0, x))$ for the decomposition of $y^\varepsilon(\cdot)$ provided by Lemma 3.4. Let (α_j, β_j) be defined by (3.53), i.e.,

$$a_j = y_1(j\bar{k}nT; 0, x) \text{ and } \beta_j = y_2(j\bar{k}nT; 0, x) \text{ for each } j \in \{0, 1, 2, \dots\}. \tag{3.123}$$

Then, from (3.122) and (3.120) (which correspond to respectively (3.47) and (3.52)), by the same arguments as those in the proof of Theorem 3.1 in Sect. 3.2.1 (noticing that $\widehat{\Phi}^\varepsilon(\cdot)$ here corresponds to $\Psi_\varepsilon(\cdot)$ there), we can reach the estimate (3.58) for the current case, i.e., for all $j = 0, 1, \dots,$

$$\begin{pmatrix} \|\alpha_{j+1}\| \\ \|\beta_{j+1}\| \end{pmatrix} \le \begin{pmatrix} e^{-\rho_1 \bar{k}nT} & e^{-\rho_2 \bar{k}nT} + \left\|\widehat{\Phi}^\varepsilon\left(\bar{k}nT\right)\right\| \\ 0 & e^{-\rho_2 \bar{k}nT} \end{pmatrix} \begin{pmatrix} \|\alpha_j\| \\ \|\beta_j\| \end{pmatrix}. \tag{3.124}$$

Here, we have used (3.120) and (3.121). Meanwhile, it follows from (3.123) and Lemma 3.4 that for all $j = 0, 1, \dots,$

$$\alpha_j = y_1^\varepsilon(j\bar{k}nT) = \mathrm{Proj}_V\left(y^\varepsilon(j\bar{k}nT)\right),\ \beta_j = y_2^\varepsilon(j\bar{k}nT) = \mathrm{Proj}_{V^\perp}\left(y^\varepsilon(j\bar{k}nT)\right). \tag{3.125}$$

Let $M_2^\varepsilon \triangleq 1 + \left\|\widehat{\Phi}^\varepsilon\left(\bar{k}nT\right)\right\|$. Then, by (3.124), one can easily check that for all j,

$$\begin{pmatrix} \|\alpha_j\| \\ \|\beta_j\| \end{pmatrix} \le \begin{pmatrix} e^{-j\rho_1 \bar{k}nT} & M_2^\varepsilon \sum_{l=0}^{j-1} e^{-(l\rho_1 + (j-1-l)\rho_2)\bar{k}nT} \\ 0 & e^{-j\rho_2 \bar{k}nT} \end{pmatrix} \begin{pmatrix} \|\alpha_0\| \\ \|\beta_0\| \end{pmatrix}. \tag{3.126}$$

Because $\rho_1 \geq \rho_2 + 1$ (see (3.121)), we find that for each $j = 1, 2, \ldots,$

$$\sum_{l=0}^{j-1} e^{-(l\rho_1+(j-1-l)\rho_2)\bar{k}nT} = e^{-(j-1)\rho_2\bar{k}nT} \sum_{l=0}^{j-1} e^{-l(\rho_1-\rho_2)\bar{k}nT}$$
$$\leq e^{-(j-1)\rho_2\bar{k}nT} \sum_{l=0}^{j-1} e^{-l\bar{k}nT} \leq e^{-(j-1)\rho_2\bar{k}nT} \sum_{l=0}^{\infty} e^{-l\bar{k}nT} = \frac{e^{\rho_2\bar{k}nT}}{1-e^{-\bar{k}nT}} e^{-j\rho_2\bar{k}nT}. \tag{3.127}$$

Let $M_3^\varepsilon \triangleq M_2^\varepsilon e^{\rho_2\bar{k}nT}/(1-e^{-\bar{k}nT})$. It follows from (3.126) and (3.127) that for all j,

$$\|y^\varepsilon(j\bar{k}nT)\| \leq \|\alpha_j\| + \|\beta_j\| \leq e^{-\rho_1 j\bar{k}nT}\|\alpha_0\| + (1+M_3^\varepsilon)e^{-\rho_2 j\bar{k}nT}\|\beta_0\|. \tag{3.128}$$

By (3.128) and by the same argument in (3.118), we find that there is a $C > 0$ satisfying

$$\|y^\varepsilon(t)\| \leq C\left(e^{-\rho_1 t}\|\alpha_0\| + e^{-\rho_2 t}\|\beta_0\|\right). \tag{3.129}$$

Because $\rho_1 \geq 1/\delta$ (see (3.121)) and $\rho_2 \geq \rho_0 - \delta$ (see (3.119)), the estimate (3.129), as well as (3.125) with $j = 0$, leads to (3.114). This completes the proof. □

3.3 Geometric Criterion

This section is devoted to introduce a geometric criterion for a T-periodic pair $[A(\cdot), B(\cdot)]$ being periodic stabilizable. We first try to figure out such a geometric condition from perspective of the LQ theory. Let $[A(\cdot), B(\cdot)]$ be a T-periodic pair with V and $\mathscr{P}$. Recalling Lemma 1.2 and Remark 1.2, we find that

$$\mathbb{R} = \mathbb{R}_1(\mathscr{P}) \oplus \mathbb{R}_2(\mathscr{P}),$$

where $\mathbb{R}_1(\mathscr{P})$ and $\mathbb{R}_2(\mathscr{P})$ are invariant subspaces of $\mathscr{P}$, with

$$\sigma(\mathscr{P}|_{\mathbb{R}_1(\mathscr{P})}) \subset \mathbb{B} \quad \text{and} \quad \sigma(\mathscr{P}|_{\mathbb{R}_2(\mathscr{P})}) \subset \mathbb{B}^c.$$

Thus, each $x \in \mathbb{R}^n$ can be expressed by $x = x_1 + x_2$, with $x_1 \in \mathbb{R}_1^n(\mathscr{P})$ and $x_2 \in \mathbb{R}_2^n(\mathscr{P})$. Let us assume that

$$\mathbb{R}_2^n(\mathscr{P}) \subseteq V. \tag{3.130}$$

Consider the LQ problem $(LQ)_{0,x}^\infty$ (with $x \in \mathbb{R}^n$), defined by (1.89) with $Q = I_n$ and $R = I_m$. From (3.130), we find that $x_2 \in V$. From this and Lemma 3.1, there is a control $u \in \mathscr{U}_{ad}$ so that $J_{0,x_2}^\infty(u) < \infty$. Write $y_1(\cdot) = y(\cdot; 0, x_1; 0)$ and $y_2(\cdot) = y(\cdot; 0, x_2, u)$. Since $x_1 \in \mathbb{R}_1^n(\mathscr{P})$, there are two positive constants C and δ so that $\|y_1(s)\| \leq Ce^{-\delta s}$ for all $s \in \mathbb{R}^+$. Then by (3.18) and (3.19), we see that

$$J_{0,x}^{\infty}(u) \leq 2\int_0^{\infty} \|y_1(s)\|^2 \mathrm{d}s + 2\int_0^{\infty} \left[\|y_2(s)\|^2 + \|u(s)\|^2\right]\mathrm{d}s$$
$$\leq \int_0^{\infty} 2C^2 e^{-2\delta s}\mathrm{d}s + 2J_{0,x_2}^{\infty}(u) < \infty.$$

From this and Theorem 1.4, we find that the pair $\left[A(\cdot), B(\cdot)\right]$ is T-periodically stabilizable. Hence, (3.130) is a sufficient condition to ensure the T-periodic stabilization of $[A(\cdot), B(\cdot)]$.

Now, a natural question is whether (3.130) is also a necessary condition to the periodic feedback stabilization of $[A(\cdot), B(\cdot)]$. The answer is positive.

Theorem 3.3 *Let $\left[A(\cdot), B(\cdot)\right]$ be a T-periodic pair with the null controllable subspace V. Then the following assertions are equivalent:*

(i) The pair $\left[A(\cdot), B(\cdot)\right]$ is T-periodically stabilizable.

(ii) The property (3.130) holds.

Proof In this proof, we will not use the above mentioned LQ theory. Let $\bar{Q}$ be the positive semi-definite matrix given by (3.15). By Lemma 3.3, we see that $V = \mathscr{N}(\bar{Q})$. Then, there is an orthogonal matrix $P = (p_1, \ldots, p_n)$, with p_j its jth column vector, so that

$$P^*\bar{Q}P = diag\left\{\mu_1, \ldots, \mu_{n-k_1}, 0, \ldots, 0\right\},$$

where $\mu_i > 0$ for $i = 1, \ldots, n - k_1$, with k_1 given by (3.22). Hence,

$$V = span\left\{p_{n-k_1+1}, \ldots, p_n\right\}. \tag{3.131}$$

Along with the invariance of V under $\mathscr{P}$, this indicates that

$$\mathscr{P}P = P\begin{pmatrix} A_1 & 0 \\ A_2 & A_3 \end{pmatrix}, \tag{3.132}$$

where $A_1 \in \mathbb{R}^{(n-k_1)\times(n-k_1)}$, $A_2 \in \mathbb{R}^{k_1\times(n-k_1)}$, $A_3 \in \mathbb{R}^{k_1\times k_1}$. Hence,

$$P^* \cdot \bar{Q}^{\dagger}\bar{Q}\mathscr{P} \cdot P = (P^*\bar{Q}^{\dagger}P)(P^*\bar{Q}P)(P^*\mathscr{P}P) = (P^*\bar{Q}P)^{\dagger}(P^*\bar{Q}P)(P^*\mathscr{P}P)$$
$$= diag\left\{\mu_1^{-1}, \ldots, \mu_{n-k_1}^{-1}, 0, \ldots, 0\right\} diag\left\{\mu_1, \ldots, \mu_{n-k_1}, 0, \ldots, 0\right\}(P^*\mathscr{P}P)$$
$$= \begin{pmatrix} I_{n-k_1} & 0 \\ 0 & 0_{k_1\times k_1} \end{pmatrix}\begin{pmatrix} A_1 & 0 \\ A_2 & A_3 \end{pmatrix} = \begin{pmatrix} A_1 & 0 \\ 0 & 0 \end{pmatrix}.$$

From the above and the equivalence of (b) and (c) in Theorem 3.1, it follows that

$$(i) \text{ in Theorem 3.3} \Leftrightarrow \sigma(A_1) \subset \mathbb{B}. \tag{3.133}$$

Since V is an invariant subspace of $\mathscr{P}$, we have that (see Lemma 1.2 and Remark 1.2)

$$V = V_1(\mathscr{P}) \bigoplus V_2(\mathscr{P}),\ \sigma\left(\mathscr{P}\big|_{V_1(\mathscr{P})}\right) \subset \mathbb{B},\ \sigma\left(\mathscr{P}\big|_{V_2(\mathscr{P})}\right) \subset \mathbb{B}^c. \tag{3.134}$$

From properties of $V_2(\mathscr{P})$ and $\mathbb{R}^n_2(\mathscr{P})$, one can easily check that

$$(ii) \text{ in Theorem 3.3} \Leftrightarrow V_1(\mathscr{P}) = \mathbb{R}^n_2(\mathscr{P}). \tag{3.135}$$

Thus, to prove $(i) \Leftrightarrow (ii)$, it suffices to verify that

$$V_2(\mathscr{P}) = \mathbb{R}^n_2(\mathscr{P}) \Leftrightarrow \sigma(A_1) \subset \mathbb{B}. \tag{3.136}$$

Now, we are on the position to prove (3.136). Write $V_1(\mathscr{P}) = span\{\eta_1, \ldots, \eta_{k_2}\}$ with the basis $\{\eta_1, \ldots, \eta_{k_2}\}$ and $V_2(\mathscr{P}) = span\{\eta_{k_2+1}, \ldots, \eta_{k_1}\}$ with the basis $\{\eta_{k_2+1}, \ldots, \eta_{k_1}\}$. Then by (3.134), there are invertible matrices $A_{31} \in \mathbb{R}^{k_2 \times k_2}$ and $A_{32} \in \mathbb{R}^{(k_1-k_2)\times(k_1-k_2)}$ so that

$$\mathscr{P}(\eta_1, \ldots, \eta_{k_1}) = (\eta_1, \ldots, \eta_{k_1}) \begin{pmatrix} A_{31} & 0 \\ 0 & A_{32} \end{pmatrix}; \tag{3.137}$$

$$\sigma(A_{31}) = \sigma\left(\mathscr{P}\big|_{V_1(\mathscr{P})}\right) \subset \mathbb{B}, \qquad \sigma(A_{32}) = \sigma\left(\mathscr{P}\big|_{V_2(\mathscr{P})}\right) \subset \mathbb{B}^c. \tag{3.138}$$

Hence, $\{p_1, \ldots, p_{n-k_1}, \eta_1, \ldots, \eta_{k_1}\}$ is a basis of $\mathbb{R}^n$. From (3.132) and (3.137), we find that

$$\begin{aligned} &\mathscr{P}(p_1, \ldots, p_{n-k_1}, \eta_1, \ldots, \eta_{k_1-k_2}, \ldots, \eta_{k_1}) \\ &= (p_1, \ldots, p_{n-k_1}, \eta_1, \ldots, \eta_{k_1-k_2}, \ldots, \eta_{k_1}) \begin{pmatrix} A_1 & 0 & 0 \\ A_{21} & A_{31} & 0 \\ A_{22} & 0 & A_{32} \end{pmatrix} \end{aligned}$$

for some $A_{21} \in \mathbb{R}^{k_2 \times (n-k_1)}$ and $A_{22} \in \mathbb{R}^{(k_1-k_2)\times(n-k_1)}$. Write

$$\hat{A}_1 = \begin{pmatrix} A_1 & 0 \\ A_{21} & A_{31} \end{pmatrix} \in \mathbb{R}^{(n-k_1+k_2)\times(n-k_1+k_2)}, \tag{3.139}$$

$$\hat{A}_2 = \left(A_{22}, 0_{(k_1-k_2)\times k_2}\right) \in \mathbb{R}^{(k_1-k_2)\times(n-k_1+k_2)}$$

and

$$\hat{P} = (p_1, \ldots, p_{n-k_1}, \eta_1, \ldots, \eta_{k_1}). \tag{3.140}$$

One can directly check that

$$\hat{P}^{-1}\mathscr{P}\hat{P} = \begin{pmatrix} \hat{A}_1 & 0 \\ \hat{A}_2 & A_{32} \end{pmatrix}. \tag{3.141}$$

The rest of the proof of (3.136) is carried out by the following two steps.

Step 1: To show that $\sigma(A_1) \subset \mathbb{B} \Rightarrow V_2(\mathscr{P}) = \mathbb{R}_2^n(\mathscr{P})$

Suppose that $\sigma(A_1) \subset \mathbb{B}$. Then, by (3.138) and (3.139), we find that $\sigma(\hat{A}_1) \subset \mathbb{B}$. This, along with (3.138), yields that $\sigma(\hat{A}_1) \bigcap \sigma(A_{32}) = \varnothing$. Thus, the Sylvster equation:

$$A_{32}X - X\hat{A}_1 = \hat{A}_2 \tag{3.142}$$

has a unique solution $\hat{X} \in \mathbb{R}^{(k_1-k_2)\times(n-k_1+k_2)}$ (see p. 131, [88]). Let

$$\widetilde{P} \triangleq \hat{P}\begin{pmatrix} I_{n-k_1+k_2} & 0 \\ -\hat{X} & I_{k_1-k_2} \end{pmatrix}. \tag{3.143}$$

From (3.141), (3.142) and (3.143), it follows that

$$\widetilde{P}^{-1}\mathscr{P}\widetilde{P} = \begin{pmatrix} \hat{A}_1 & 0 \\ 0 & A_{32} \end{pmatrix}. \tag{3.144}$$

Write $\widetilde{p}_j$ for the j-th column vector of $\widetilde{P}$. Let

$$Y_1 = span\left\{\widetilde{p}_1, \widetilde{p}_2, \ldots, \widetilde{p}_{n-k_1+k_2}\right\}; \quad Y_2 = span\left\{\widetilde{p}_{n-k_1+k_2+1}, \widetilde{p}_{n-k_1+k_2+2}, \ldots, \widetilde{p}_n\right\}. \tag{3.145}$$

By (3.144), they are invariant subspaces of $\mathscr{P}$ and satisfy that

$$Y_1 \oplus Y_2 = \mathbb{R}^n, \ \sigma(\mathscr{P}\big|_{Y_1}) = \sigma(\hat{A}_1) \subset \mathbb{B}, \ \sigma(\mathscr{P}\big|_{Y_2}) = \sigma(A_{32}) \subset \mathbb{B}^c.$$

Two facts are given in order: First, it follows by Lemma 1.2 that $Y_2 = \mathbb{R}_2^n(\mathscr{P})$. Second, it follows by (3.143) and (3.140) that

$$\widetilde{p}_{n-k_1+k_2+j} = \eta_{k_2+j} \quad \text{for all} \quad j = 1, \ldots, k_1 - k_2.$$

Since

$$V_2(\mathscr{P}) = span\{\eta_{k_2+1}, \ldots, \eta_{k_1}\},$$

the above two facts, together with (3.145), yield that $\mathbb{R}_2^n(\mathscr{P}) = V_2(\mathscr{P})$.

Step 2: To show that $V_2(\mathscr{P}) = \mathbb{R}_2^n(\mathscr{P}) \Rightarrow \sigma(A_1) \subset \mathbb{B}$

Suppose that $V_2(\mathscr{P}) = \mathbb{R}_2^n(\mathscr{P})$. Then we see that $k_1 - k_2 = n - k_3$, where k_3 is given by (3.22). By Lemma (1.3), it follows that $V_1(\mathscr{P}) \subseteq \mathbb{R}_1^n(\mathscr{P})$. Thus we can write

$$\mathbb{R}_1^n(\mathscr{P}) = span\{\beta_1, \ldots, \beta_{n-k_1}, \eta_1, \ldots, \eta_{k_2}\}. \tag{3.146}$$

Hence, we have that

$$\mathbb{R}_1^n(\mathscr{P}) \bigoplus V_2(\mathscr{P}) = \mathbb{R}_1^n(\mathscr{P}) \bigoplus \mathbb{R}_2^n(\mathscr{P}) = \mathbb{R}^n.$$

Because $V_2(\mathscr{P}) = span\{\eta_{k_2+1}, \ldots, \eta_{k_1}\}$, the above, along with (3.146) and (3.140), yields

$$\left(\beta_1, \ldots, \beta_{n-k_1}, \eta_1, \ldots, \eta_{k_1}\right) = \left(p_1, \ldots, p_{n-k_1}, \eta_1, \ldots, \eta_{k_1}\right) \begin{pmatrix} \hat{A}_3 & 0 \\ \hat{A}_4 & I_{k_1} \end{pmatrix}, \tag{3.147}$$

where $\hat{A}_3 \in \mathbb{R}^{(n-k_1)\times k_1}$ is invertible and $\hat{A}_4 \in \mathbb{R}^{k_1\times(n-k_1)}$. Since $\mathbb{R}_1^n(\mathscr{P})$ is invariant under $\mathscr{P}$ and $\sigma(\mathscr{P}|_{\mathbb{R}_1^n(\mathscr{P})}) \subset \mathbb{B}$, by (3.137), there are matrices $\hat{A}_6 \in \mathbb{R}^{k_2\times(n-k_1)}$ and $\hat{A}_5 \in \mathbb{R}^{(n-k_1)\times(n-k_1)}$, with $\sigma(\hat{A}_5) \subset \mathbb{B} \setminus \{0\}$, so that

$$\begin{aligned} &\mathscr{P}\left(\beta_1, \ldots, \beta_{n-k_1}, \eta_1, \ldots, \eta_{k_2}, \eta_{k_2+1} \ldots, \eta_{k_1}\right) \\ &= \left(\beta_1, \ldots, \beta_{k_2}, \eta_1, \ldots, \eta_{k_2}, \eta_{k_2+1} \ldots, \eta_{k_1}\right) \begin{pmatrix} \hat{A}_5 & 0 & 0 \\ \hat{A}_6 & A_{31} & 0 \\ 0 & 0 & A_{32} \end{pmatrix}. \end{aligned} \tag{3.148}$$

Write

$$\hat{A}_7 = \begin{pmatrix} \hat{A}_6 \\ 0_{(k_1-k_2)\times(n-k_1)} \end{pmatrix} \in \mathbb{R}^{k_1\times(n-k_1)}, \qquad \hat{A}_8 = \begin{pmatrix} A_{31} & 0 \\ 0 & A_{32} \end{pmatrix} \in \mathbb{R}^{k_1\times k_1}. \tag{3.149}$$

From (3.140), (3.147), (3.148) and (3.149), we can directly check that

$$\mathscr{P}\hat{P} = \hat{P}\begin{pmatrix} \hat{A}_3\hat{A}_5\hat{A}_3^{-1} & 0_{(n-k_1)\times k_1} \\ * & * \end{pmatrix}. \tag{3.150}$$

Because both $\{p_{n-k_1+1}, \ldots, p_n\}$ and $\{\eta_1, \ldots, \eta_{k_1}\}$ are base of V, there is an invertible matrix $A_9 \in \mathbb{R}^{k_1\times k_1}$ so that $\hat{P} = P \cdot diag\{I_{n-k_1}, A_9\}$. Then, it follows from (3.150) that

$$P^{-1}\mathscr{P}P = \begin{pmatrix} \hat{A}_3\hat{A}_5\hat{A}_3^{-1} & 0 \\ * & * \end{pmatrix}.$$

Combined with (3.132), the above shows that $A_1 = \hat{A}_3\hat{A}_5\hat{A}_3^{-1}$. Thus, $\sigma(A_1) = \sigma(\hat{A}_5) \subset \mathbb{B}$.

In summary, we complete the proof of Theorem 3.3. $\square$

Miscellaneous Notes

There have been studies on periodic stabilization criteria for linear periodic systems. In [45], the following criterion was established (see Theorem 2 in [45]): T-periodic stabilization $\Leftrightarrow$ H-stabilization. Here, a T-periodic pair $\big[A(\cdot), B(\cdot)\big]$ is said to be H-stabilizable, if for each $\lambda \in \sigma(\mathscr{P})$ with $|\lambda| \geq 1$, it holds that

$$\eta = 0, \quad \text{when } \mathscr{P}^*\eta = \lambda\eta \text{ and } B^*(t)(\widehat{\Phi}(t)^*)^{-1}\eta = 0 \text{ for a.e. } t \in [0, T]. \tag{3.151}$$

The property (3.151) is a kind of unique continuation property for eigenfunctions of $\mathscr{P}^*$ corresponding to $\lambda \in \sigma(\mathscr{P})$ with $|\lambda| \geq 1$. There is a similar version of this kind of unique continuation property in infinite dimensional systems (see [7]). With the aid of the above-mentioned criterion, the authors of [45] built up a T-periodic feedback stabilization law via the solution of the following T-periodic matrix Riccati equation

$$\dot{Q} + A^*Q + QA + H^*H - QBB^*Q = 0; \ Q(t) = Q(t+T), \quad t \in \mathbb{R}^+, \tag{3.152}$$

where $H(\cdot)$ is a T-periodic matrix-valued function such that the system $[A(\cdot)^*, H(\cdot)^*]$ is H-stabilizable. It was proved in [44] that when both $\big[A(\cdot), B(\cdot)\big]$ and $[A(\cdot)^*, H(\cdot)^*]$ are H-stabilizable, the Eq. (3.152) admits a unique positive semi-definite matrix-valued T-periodic solution (see [44]). Furthermore, $K(\cdot) = -B(\cdot)^*Q(\cdot)$ is a T-periodical feedback stabilization law. The study in [45] was partially motivated by [18], where the author proved that a T-periodic pair $[A(\cdot), B(\cdot)]$ is controllable if and only if it is controllable over $[0, nT]$. This also hints us to define the concept of nT-periodic stabilization and to build up a nT-periodic stabilization law.

The periodic stabilization criteria introduced in this chapter differ from that in [45]. The equivalent condition (c) in Theorem 3.1 is a natural extension of Kalman's rank condition (see Remark 3.3). The matrix X in Theorem 3.1 can be explicitly structured. When $[A(\cdot), B(\cdot)]$ is T-periodically stabilizable, one can use a very similar method as that used in the proof of Theorem 3.2 to derive a similar estimate to (3.114) for solutions $y^{\varepsilon}(\cdot)$ to Eq. (3.6) with $K(\cdot) = K^{\varepsilon}(\cdot)$ where $K^{\varepsilon}(\cdot)$ is defined by (3.28) with X given by Theorem 3.1. The byproduct that $V = \mathscr{N}(\bar{Q})$ (see Lemma 3.3) gives a connection between the null-controllable subspace V of a T-periodic pair $[A(\cdot), B(\cdot)]$ and the corresponding matrix $\bar{Q}$.

R. Brockett formulated the following problem in [17]: *What are the conditions on a triple* (A, B, C) *(with* $A \in \mathbb{R}^{n\times n}$, $B \in \mathbb{R}^{n\times m}$ *and* $C \in \mathbb{R}^{p\times n}$*) ensuring the existence of a periodic* $K(\cdot)$ *(with* $K(t) \in \mathbb{R}^{m\times p}$*) such that the system* $\dot{y}(t) = Ay(t) + BK(t)Cy(t)$ *is asymptotically stable?* After the Brockett problem, it was pointed out in [58] that *however, the stabilizaiton of the above system by a constant* *matrix K is a classical problem in the control theory, from this point of view, the Brockett problem can be reformulated as: Can the time periodic matrices $K(t)$ aid in the stabilization?* Furthermore, the positive answer for the reformulated Brockett problem (at least for the case where $n = 2$) was given in [58]. The connections of our Theorem 3.1 and the reformulated Brockett problem are as follows. By Theorem 3.1, we will find that (see Remark 3.3) when $\left[A(\cdot), B(\cdot)\right] = [A, B]$ is time invariant, (3.1) is T-periodically stabilizable for some $T > 0$ if and only if (3.1) is T-periodically stabilizable for any $T > 0$ if and only if (3.1) is feedback stabilizable by a constant matrix. Hence, the time periodic matrices $K(t)$ will not aid in the stabilization of any triple (A, B, C) with Rank$C = n$, i.e., the reformulated Brockett problem has the positive answer only if Rank$C < n$.

The material of this chapter is adapted from [92].

Chapter 4
Design of Simple Control Machines

Abstract Let $A(\cdot) \in L^\infty(\mathbb{R}^+; \mathbb{R}^{n\times n})$ be T-periodic. In the case that $\sigma(\mathscr{P}) \subseteq \mathbb{B}$ (where $\mathscr{P}$ is given by (3.5)), we need to do nothing from perspective of the periodic stabilization. If $\sigma(\mathscr{P}) \nsubseteq \mathbb{B}$, we will find a T-periodic control machine $B(\cdot) \in L^\infty(\mathbb{R}^+; \mathbb{R}^{n\times m})$ so that $[A(\cdot), B(\cdot)]$ is linear T-periodic feedback stabilizable (LPFS, for short). Among all such control machines, how to choose a simple one? To answer this question, we should first explain what means *simple one*. From different perspectives of applications, one can give different definitions of simple machines. We will define two kinds of simple control machines and provide the ways to design the corresponding simple control machines. When design them, Theorem 3.3 will be used. Throughout this chapter, we focus ourself on those T-periodic $A(\cdot)$ with $\sigma(\mathscr{P}) \nsubseteq \mathbb{B}$.

Keywords Periodic Equations · Stabilization · Simple Control Machines · ODE

The first kind of simple control machines is defined as follows: For each T-periodic $A(\cdot) \in L^\infty(\mathbb{R}^+; \mathbb{R}^n)$, we define

$$\mathscr{C}\mathscr{B}^1_{A(\cdot)} \triangleq \left\{B(\cdot) \in L^\infty(\mathbb{R}^+; \mathbb{R}^{n\times m}) \;\middle|\; m \in \mathbb{N}, [A(\cdot), B(\cdot)] \text{ is LPFS}\right\}. \tag{4.1}$$

For each $B(\cdot) \in \mathscr{C}\mathscr{B}^1_{A(\cdot)}$, denote by $N(B(\cdot))$ the number of columns of $B(\cdot)$. Set

$$N\left(\mathscr{C}\mathscr{B}^1_{A(\cdot)}\right) \triangleq \min\left\{N(B(\cdot)) \;\middle|\; B(\cdot) \in \mathscr{C}\mathscr{B}^1_{A(\cdot)}\right\}. \tag{4.2}$$

The first kind of simple control machines w.r.t. $A(\cdot)$ is defined to be such $B(\cdot) \in \mathscr{C}\mathscr{B}^1_{A(\cdot)}$ with $m = N\left(\mathscr{C}\mathscr{B}^1_{A(\cdot)}\right)$.

The second kind of simple control machines is defined as follows: For each T-periodic $A(\cdot) \in L^\infty(\mathbb{R}^+; \mathbb{R}^n)$, define

$$\mathscr{C}\mathscr{B}_{A(\cdot)} \triangleq \left\{\hat{B} \in \mathbb{R}^{n\times m} \;\middle|\; m \in \mathbb{N}, [A(\cdot), \hat{B}] \text{ is LTPFS}\right\}. \tag{4.3}$$

G. Wang and Y. Xu, *Periodic Feedback Stabilization for Linear Periodic Evolution Equations*, SpringerBriefs in Mathematics,
DOI 10.1007/978-3-319-49238-4_4

For each $\hat{B} \in \mathscr{CB}_{A(\cdot)}$, denote by $M(\hat{B})$ the number of columns of $\hat{B}$. Set

$$M\left(\mathscr{CB}_{A(\cdot)}\right) \triangleq \min\left\{M(\hat{B}) \mid \hat{B} \in \mathscr{CB}_{A(\cdot)}\right\}. \tag{4.4}$$

The second kind of simple control machines w.r.t. $A(\cdot)$ is defined to be such $\hat{B} \in \mathscr{CB}_{A(\cdot)}$ with $m = M\left(\mathscr{CB}_{A(\cdot)}\right)$.

4.1 The First Kind of Simple Control Machines

This section is devoted to the studies of the first kind of simple control machines. First of all we introduce the following lemma on the Floquet theory (see [35, 71]). We omit its proof.

Lemma 4.1 *Let $A(\cdot) \in L^\infty(\mathbb{R}^+; \mathbb{R}^n)$ be T-periodic. Then there is a T-periodic transform $\mathscr{T}(\cdot) \in C^1(\mathbb{R}; \mathbb{R}^{n\times n})$, with non-singular $\mathscr{T}(t)$ for each t, (which is called the Lyapunov transform associated with $A(\cdot)$,) so that the time varying system $\dot{x}(t) = A(t)x(t)$ can be transformed into an autonomous system with respect to z by the transformation $x = \mathscr{T}(t)z$.*

Theorem 4.1 *Let $A(\cdot) \in L^\infty(\mathbb{R}^+; \mathbb{R}^n)$ be T-periodic. Then there is a T-periodic function $B(\cdot)$ in $L^\infty(\mathbb{R}^+; \mathbb{R}^{n\times 1})$ so that $[A(\cdot), B(\cdot)]$ is periodically stabilizable, i.e., $N\left(\mathscr{CB}^1_{A(\cdot)}\right) = 1$.*

Proof Let $\mathscr{T}(\cdot)$ be the Lyapunov transform given by Lemma 4.1. Then we have that

$$\mathscr{T}^{-1}(t)\left(A(t)\mathscr{T}(t) - \dot{\mathscr{T}}(t)\right) \equiv \hat{A}, \quad t \geq 0 \tag{4.5}$$

for some $\hat{A} \in \mathbb{R}^{n\times n}$. Let $\{e_1, \ldots, e_n\}$ be the standard basis of $\mathbb{R}^n$. Write

$$b_k = e^{(kT/n)\hat{A}} e_{k+1}, \quad k = 0, \ldots, n-1. \tag{4.6}$$

Define $\hat{B}(\cdot)$ by

$$\hat{B}(t) = \sum_{k=0}^{n-1} \chi_{[kT/n,(k+1)T/n)} b_k \ \text{ for } t \in [0, T); \quad \hat{B}(t) = \hat{B}(t-T) \ \text{ for } t \geq T. \tag{4.7}$$

Write $B(\cdot) = \mathscr{T}(\cdot)\hat{B}(\cdot)$. Clearly, $B(\cdot)$ is a T-periodic function in $L^\infty(\mathbb{R}^+; \mathbb{R}^{n\times 1})$. According to Lemma 4.1 and (4.5), the following two controlled systems are equivalent:

$$\dot{x}(t) = A(t)x(t) + B(t)u(t), \quad t \geq 0 \tag{4.8}$$

and

$$\dot{z}(t) = \hat{A}z(t) + \hat{B}(t)u(t), \quad t \geq 0. \tag{4.9}$$

For each $t_0 \geq 0$, $z_0 \in \mathbb{R}^n$ and each control u, we denote by $z(\cdot; t_0, z_0, u)$ the solution of (4.9) with the initial condition that $z(t_0) = z_0$.

We claim that $[A(\cdot), B(\cdot)]$ is periodically stabilizable. By Theorem 3.3 and the equivalence of systems (4.8) and (4.9), it suffices to show the null controllability of System (4.9) over $[0, T]$. For this purpose, we write $z_k(\cdot; z_k, u)$ for the solution to the system:

$$\dot{z}(t) = \hat{A}z(t) + b_k u(t), \quad t \in [kT/n, (k+1)T/n], \quad z(kT/n) = z_k. \tag{4.10}$$

By Lemma 3.3.2 in [86] (see p. 91 in [86]), $span\{b_k, \hat{A}b_k, \ldots, \hat{A}^{n-1b_k}\}$ is a controllable subspace of (4.10). In particular, for each $k \in \{0, 1, 2, \ldots, n-1\}$, there is a control $u_k \in L^2(\mathbb{R}^+; \mathbb{R}^1)$ so that $z_k((k+1)T/n; b_k, u_k) = 0$. This, along with definitions of $z(\cdot; t_0, z_0, u)$ and $z_k(\cdot; z_k, u)$, (4.6), (4.7), (4.9) and (4.10), yields that for all $k \in \{0, 1, 2, \ldots, n-1\}$,

$$\begin{aligned} z(T; 0, e_{k+1}, \chi_{[kT/n,(k+1)T/n)}u_k) &= z(T; kT/n, b_k, \chi_{[kT/n,(k+1)T/n)}u_k) \\ &= z(T; (k+1)T/n, 0, \chi_{[kT/n,(k+1)T/n)}u_k) = 0, \end{aligned}$$

which leads to the null controllability of System (4.9) and completes the proof. □

Remark 4.1 In the proof of Theorem 4.1, we have provided a way to construct a $B(\cdot) \in \mathscr{CB}^1_{A(\cdot)}$, with $m = 1$, through utilizing the Lyapunov transform.

4.2 The Second Kind of Simple Control Machines–General Case

This section is devoted to the studies of the second kind of simple control machines for the general case where $A(\cdot)$ is time varying. First of all, we recall (3.22) for the definitions of k_i with $i = 1, 2, 3$. The following proposition is a consequence of Theorem 3.3. It is the base of studies on the second kind of simple control machines.

Proposition 4.1 *Let $[A(\cdot), B(\cdot)]$ be a T-periodic pair with $\mathscr{P}$ and V. Let $V_2(\mathscr{P})$, $\mathbb{R}^n_2(\mathscr{P})$ and $\mathbb{R}^n_1(\mathscr{P})$ be given by (1.47). Then, the following statements are equivalent: (a) $\big[A(\cdot), B(\cdot)\big]$ is LPFS; (b) $\mathbb{R}^n_2(\mathscr{P}) \subseteq V$; (c) $V + \mathbb{R}^n_1(\mathscr{P}) = \mathbb{R}^n$; (d) $V_2(\mathscr{P}) = \mathbb{R}^n_2(\mathscr{P})$.*

Proof First, it follows from Theorem 3.3 that $(a) \Leftrightarrow (b)$. Next, we show $(c) \Rightarrow (d) \Rightarrow (b)$. Suppose that (c) stands. Since $V = V_1(\mathscr{P}) \bigoplus V_2(\mathscr{P})$, we have that

$$\left(V_1(\mathscr{P}) + \mathbb{R}_1^n(\mathscr{P})\right) + V_2(\mathscr{P}) = \mathbb{R}^n. \tag{4.11}$$

It follows from Lemma 1.3 that

$$V_1(\mathscr{P}) + \mathbb{R}_1^n(\mathscr{P}) = \mathbb{R}_1^n(\mathscr{P}) \text{ and } V_2(\mathscr{P}) \subseteq \mathbb{R}_2^n(\mathscr{P}).$$

Thus, (d) holds. Since $V = V_1(\mathscr{P}) \bigoplus V_2(\mathscr{P})$, we find that $(d) \Rightarrow (b)$. Finally, we show that $(b) \Rightarrow (c)$. By (b), we see that

$$\mathbb{R}^n \supseteq V + \mathbb{R}_1^n(\mathscr{P}) \supseteq \mathbb{R}_2^n(\mathscr{P}) + \mathbb{R}_1^n(\mathscr{P}) = \mathbb{R}^n,$$

which leads to (c). This completes the proof. □

For each T-periodic $A(\cdot) \in L^\infty(\mathbb{R}^+; \mathbb{R}^n)$ with the map $\mathscr{P}$, one can find an invertible matrix $\hat{Q}$ so that

$$\hat{Q}^{-1}\mathscr{P}\hat{Q} = \begin{pmatrix} \mathscr{P}_1 & \\ & \mathscr{P}_2 \end{pmatrix}, \tag{4.12}$$

where $\mathscr{P}_1$ and $\mathscr{P}_2$ are real square matrices with $\sigma(\mathscr{P}_1)$ in $\mathbb{B}$ and $\sigma(\mathscr{P}_2)$ in $\mathbb{B}^c$ respectively. Write q_j for the jth column vector in $\hat{Q}$. Then, by (3.22),

$$k_3 \triangleq \dim\left(\mathbb{R}_1^n(\mathscr{P})\right) = \dim(\mathscr{P}_1) \tag{4.13}$$

and

$$\mathbb{R}_1^n(\mathscr{P}) = span\{q_1, \ldots, q_{k_3}\}. \tag{4.14}$$

The next Corollary 4.1 is a consequence of Proposition 4.1.

Corollary 4.1 *Let $[A(\cdot), B(\cdot)]$ be a T-periodic pair, with $\mathscr{P}$, V, $\hat{Q}$ and k_3 given by (3.5), (3.7), (4.12) and (4.13), respectively. Then, $[A(\cdot), B(\cdot)]$ is LPFS if and only if*

$$\left(0_{(n-k_3)\times k_3}, I_{n-k_3}\right)\ \hat{Q}^{-1}V = \mathbb{R}^{n-k_3}. \tag{4.15}$$

Proof By (3.22), we have $\dim V = k_1$. Thus, we can write

$$V = span\{\eta_1, \eta_2, \ldots, \eta_{k_1}\} \text{ with the basis } \{\eta_1, \eta_2, \ldots, \eta_{k_1}\}. \tag{4.16}$$

Clearly, there are matrices $C_1 \in \mathbb{R}^{k_3\times k_1}$, $C_2 \in \mathbb{R}^{(n-k_3)\times k_1}$ so that

$$\begin{aligned} &(q_1, q_2, \ldots, q_{k_3}, \eta_1, \eta_2, \ldots, \eta_{k_1}) \\ &= (q_1, q_2, \ldots, q_{k_3}, q_{k_3+1}, \ldots, q_n)\begin{pmatrix} I_{k_3} & C_1 \\ & C_2 \end{pmatrix}. \end{aligned} \tag{4.17}$$

From (4.16) and (4.17), we see that

$$\begin{aligned}
&\left(0_{(n-k_3)\times k_3}, I_{n-k_3}\right) \hat{Q}^{-1} V \\
&= \left(0_{(n-k_3)\times k_3}, I_{n-k_3}\right) \hat{Q}^{-1} span\left\{\eta_1, \eta_2, \ldots, \eta_{k_1}\right\} \\
&= \mathscr{R}\left(\left(0_{(n-k_3)\times k_3}, I_{n-k_3}\right)\begin{pmatrix} C_1 \\ C_2 \end{pmatrix}\right) = \mathscr{R}(C_2).
\end{aligned} \tag{4.18}$$

Since $\{q_1, \ldots, q_n\}$ is a basis of $\mathbb{R}^n$, from Proposition 4.1, (4.14), (4.16), (4.17) and (4.18), we can easily check that

$$\mathbb{R}_2^n(\mathscr{P}) \subseteq V \Leftrightarrow \left(0_{(n-k_3)\times k_3}, I_{n-k_3}\right) \hat{Q}^{-1} V = \mathbb{R}^{n-k_3}. \tag{4.19}$$

From (4.19) and Proposition 4.1, it follows that $[A(\cdot), B(\cdot)]$ is T-periodically stabilizable if and only if (4.15) holds. This completes the proof. □

Proposition 4.2 *For each T-periodic $A(\cdot)$ in $L^\infty(\mathbb{R}^+; \mathbb{R}^n)$, the $n \times n$ identity matrix I_n belongs to $\mathscr{CB}_{A(\cdot)}$. Consequently, $\mathscr{CB}_{A(\cdot)} \neq \emptyset$ and $M\left(\mathscr{CB}_{A(\cdot)}\right) \leq n$.*

Proof By Theorem 3.3, it suffices to show that $[A(\cdot), I_n]$ is null controllable over $[0, T]$. It is well known that the later is equivalent to the non-singularity of the Gramian:

$$G \triangleq \int_0^T \widehat{\Phi}^{-1}(t)(\widehat{\Phi}^{-1}(t))^* \mathrm{d}t.$$

However, the above matrix is clearly invertible in this case. This completes the proof. □

In what follows, we arbitrarily fix a T-periodic $A(\cdot) \in L^\infty(\mathbb{R}^+; \mathbb{R}^n)$ with the fundamental solution $\widehat{\Phi}(\cdot)$. Let

$$\widehat{\Phi}([0, nT]) \triangleq span\{\widehat{\Phi}^{-1}(s) \mid s \in [0, nT]\} \text{ and } \bar{k} \triangleq \dim(\widehat{\Phi}([0, nT])). \tag{4.20}$$

The following proposition provides a way to determine $\bar{k}$.

Proposition 4.3 *It holds that*

$$dim(\widehat{\Phi}([0, nT])) \triangleq \bar{k} = Rank\left(\int_0^{nT} vec(\widehat{\Phi}^{-1}(t))\left(vec(\widehat{\Phi}^{-1}(t))\right)^* \mathrm{d}t\right), \tag{4.21}$$

where $vec(\cdot) : \mathbb{R}^{n\times n} \mapsto \mathbb{R}^{n^2\times 1}$ is defined by

$$vec(D) = (d_{11}, \ldots, d_{n1}, d_{12}, \ldots, d_{n2}, \ldots, d_{1n}, \ldots, d_{nn})^*$$

for each $D = (d_{ij})_{n\times n}$.

Proof Let $\{\widehat{\Phi}_1, \ldots, \widehat{\Phi}_{\bar{k}}\}$ satisfy that

$$span\left\{\widehat{\Phi}_1, \ldots, \widehat{\Phi}_{\bar{k}}\right\} = span\{\widehat{\Phi}^{-1}(s) \mid s \in [0, nT]\}.$$

By the continuity of $\widehat{\Phi}^{-1}(\cdot)$, there is a $\mathbb{R}^{\bar{k}}$-valued continuous functions $c(\cdot) = (c_1(\cdot), \ldots, c_{\bar{k}}(\cdot))^*$ so that

$$vec(\widehat{\Phi}^{-1}(s)) = \sum_{j=1}^{\bar{k}} c_j(s) vec(\widehat{\Phi}_j) = \big(vec(\widehat{\Phi}_1), \ldots, vec(\widehat{\Phi}_{\bar{k}})\big)c(s). \tag{4.22}$$

By (4.22), one can easily check that

$$\begin{aligned}&\int_0^{nT} vec(\widehat{\Phi}^{-1}(s))\big(vec(\widehat{\Phi}^{-1}(s))\big)^* \mathrm{d}s \\ &= \big(vec(\widehat{\Phi}_1), \ldots, vec(\widehat{\Phi}_{\bar{k}})\big) \int_0^{nT} c(s)c(s)^* \mathrm{d}s \big(vec(\widehat{\Phi}_1), \ldots, vec(\widehat{\Phi}_{\bar{k}})\big)^*.\end{aligned}$$

This implies

$$\begin{aligned}&Rank\left(\int_0^{nT} vec(\widehat{\Phi}^{-1}(s))\big(vec(\widehat{\Phi}^{-1}(s))\big)^* \mathrm{d}s\right) \\ &= Rank\left(\int_0^{nT} c(s)c(s)^* \mathrm{d}s\right) \le \bar{k}.\end{aligned}$$

The rest is to show that the $\bar{k} \times \bar{k}$ matrix $\left(\int_0^{nT} c(s)c(s)^* \mathrm{d}s\right)$ is non-singular. In fact, when $\xi \in \mathbb{R}^{\bar{k}}$ verifies that

$$\xi^* \left(\int_0^{nT} c(s)c^*(s)\mathrm{d}s\right) \xi = 0,$$

we have that

$$c(s)^*\xi \equiv 0 \;\text{ for all }\; s \in [0, nT]. \tag{4.23}$$

On the other hand, because of (4.20), there are $s_1, \ldots, s_{\bar{k}} \in [0, nT]$ so that

$$\big(vec(\widehat{\Phi}^{-1}(s_1)), \ldots, vec(\widehat{\Phi}^{-1}(s_{\bar{k}}))\big) = \big(vec(\widehat{\Phi}_1), \ldots, vec(\widehat{\Phi}_{\bar{k}})\big)C, \tag{4.24}$$

(where C is an invertible $\bar{k} \times \bar{k}$ real matrix), and so that

$$vec(\widehat{\Phi}^{-1}(s)) = \big(vec(\widehat{\Phi}^{-1}(s_1)), \ldots, vec(\widehat{\Phi}^{-1}(s_{\bar{k}}))\big)\,\widetilde{c}(s) \tag{4.25}$$

for each $s \in [0, nT]$, where $\widetilde{c}(s) = (\widetilde{c}_1(s), \ldots, \widetilde{c}_{\bar{k}}(s))^*$, with $\widetilde{c}_i(\cdot)$ a function from $[0, nT]$ to $\mathbb{R}$. It follows from (4.22), (4.24) and (4.25) that

$$c(s) = C\widetilde{c}(s) \;\text{ for each }\; s \in [0, nT]. \tag{4.26}$$

From (4.25), we also find that

$$\widetilde{c}(s_j) = e_j \quad \text{for all} \quad j = 1, \ldots, \bar{k}, \tag{4.27}$$

where $\{e_1, \ldots, e_{\bar{k}}\}$ is the standard basis of $\mathbb{R}^{\bar{k}}$. Finally, from (4.23), (4.26) and (4.27),

$$C^*\xi = I_{\bar{k}} C^*\xi = \begin{pmatrix} \widetilde{c}(s_1)^* \\ \cdots \\ \widetilde{c}(s_{\bar{k}})^* \end{pmatrix} C^*\xi = \begin{pmatrix} \widetilde{c}(s_1)^* C^* \\ \cdots \\ \widetilde{c}(s_{\bar{k}})^* C^* \end{pmatrix} \xi = \begin{pmatrix} c(s_1)^* \\ \cdots \\ c(s_{\bar{k}})^* \end{pmatrix} \xi = 0.$$

which, together with the non-singularity of C, yields that $\xi = 0$. This ends the proof. □

Definition 4.1 A family of matrices $\{\widehat{\Phi}_i\}_{i=1}^{\hat{k}}$ in $\mathbb{R}^{n\times n}$, with $\hat{k} \geq \bar{k}$ (where $\bar{k}$ is given by (4.21)), is called a family generating $\widehat{\Phi}([0, nT])$ if

$$span\left\{\widehat{\Phi}_1, \ldots, \widehat{\Phi}_{\hat{k}}\right\} = span\{\widehat{\Phi}^{-1}(s) \mid s \in [0, nT]\}. \tag{4.28}$$

In general, a family generating $\widehat{\Phi}([0, nT])$ is not necessarily linearly independent. However, when $\hat{k} = \bar{k}$, it is linearly independent.

Definition 4.2 Let $\{\widehat{\Phi}_i\}_{i=1}^{\hat{k}}$ be a family generating $\widehat{\Phi}([0, nT])$. A family of matrices $\{\bar{A}_i\}_{i=1}^{\hat{k}}$ is called the family affiliated to $\{\widehat{\Phi}_i\}_{i=1}^{\hat{k}}$ if

$$\bar{A}_i = \left[\left(0_{(n-k_3)\times k_3}, I_{n-k_3}\right) \hat{Q}^{-1}\widehat{\Phi}_i\right] \in \mathbb{R}^{(n-k_3)\times n} \quad \text{for each} \quad i = 1, \ldots, \hat{k}, \tag{4.29}$$

where $\hat{Q}$ is given by (4.12).

Lemma 4.2 *Let $B \in \mathbb{R}^{n\times m}$ and assume $\{\widehat{\Phi}_i\}_{i=1}^{\hat{k}}$ is a family generating $\widehat{\Phi}([0, nT])$. Then*

$$V_{[A(\cdot),B]} = span\left\{\widehat{\Phi}_i Bv \mid 1 \leq i \leq \hat{k}, v \in \mathbb{R}^m\right\}. \tag{4.30}$$

Proof By (4.28), it suffices to show

$$V_{[A(\cdot),B]} = span\left\{\widehat{\Phi}^{-1}(s)Bv \mid s \in [0, nT], v \in \mathbb{R}^m\right\}. \tag{4.31}$$

From (3.9) and (3.11) in Lemma 3.2, we find that

$$V_{[A(\cdot),B]} = \left\{\int_0^{nT} \widehat{\Phi}^{-1}(t)Bu(t)\mathrm{d}t \mid u(\cdot) \in L^2(0, nT; \mathbb{R}^m)\right\}. \tag{4.32}$$

For each $s \in [0, nT)$ and $v \in \mathbb{R}^m$, we let

$$x^\varepsilon \triangleq \frac{1}{\varepsilon} \int_0^{nT} \widehat{\Phi}^{-1}(t)B\chi_{[s,s+\varepsilon]}(t)v\mathrm{d}t, \ \varepsilon \in (0, nT - s).$$

By (4.32), x^ε belongs to $V_{[A(\cdot),B]}$ which is closed. Thus, we see that

$$\widehat{\Phi}^{-1}(s)Bv = \lim_{\varepsilon\to 0} x^\varepsilon \in V_{[A(\cdot),B]}.$$

Similarly, we can prove that $\widehat{\Phi}^{-1}(nT)Bv \in V_{[A(\cdot),B]}$ for each $v \in \mathbb{R}^m$. Hence, the space on the right hand side of (4.31) is a subspace of $V_{[A(\cdot),B]}$.

Conversely, given $u(\cdot) \in L^2(0, nT; \mathbb{R}^m)$, there is a sequence of step functions on $[0, nT]$, denoted by $\{u_k(\cdot)\}$, so that it converges to $u(\cdot)$ in $L^2(0, nT; \mathbb{R}^m)$. One can easily check that each $\int_0^{nT} \widehat{\Phi}^{-1}(t)Bu_k(t)dt$ belongs to the space on the right hand side of (4.31). Thus,

$$\int_0^{nT} \widehat{\Phi}^{-1}(t)Bu(t)\mathrm{d}t = \lim_{k\to\infty} \int_0^{nT} \widehat{\Phi}^{-1}(t)Bu_k(t)\mathrm{d}t$$

is in the space on the right hand side of (4.31), which, along with (4.32), indicates that $V_{[A(\cdot),B]}$ is a subset of the space on the right side of (4.31). This completes the proof. □

Proposition 4.4 *Let* $\{\widehat{\Phi}_i\}_{i=1}^{\hat{k}}$ *be a family generating* $\widehat{\Phi}([0, nT])$, *with its affiliated family* $\{\bar{A}_i\}_{i=1}^{\hat{k}}$ *(see (4.29)). Then*

$$B \in \mathscr{CB}_{A(\cdot)} \Leftrightarrow \sum_{i=1}^{\hat{k}} \bar{A}_i \mathscr{R}(B) = \mathbb{R}^{n-k_3}. \tag{4.33}$$

Proof From (4.3) and Corollary 4.1, we see that $B \in \mathscr{CB}_{A(\cdot)} \Leftrightarrow$ (4.15). By Lemma 4.2, (4.15) is equivalent to

$$\left(0_{(n-k_3)\times k_3}, I_{n-k_3}\right) \hat{Q}^{-1} span\{\widehat{\Phi}_i Bv \mid 1 \le i \le \hat{k}, v \in \mathbb{R}^m\} = \mathbb{R}^{n-k_3}. \tag{4.34}$$

By (4.29), the left side of (4.34) is the space: $span\{A_i Bv \mid 1 \le i \le \hat{k}, v \in \mathbb{R}^m\}$. Meanwhile, it is clear that

$$\sum_{i=1}^{\hat{k}} \bar{A}_i \mathscr{R}(B) = span\{\bar{A}_i Bv \mid 1 \le i \le \hat{k}, v \in \mathbb{R}^n\}.$$

Hence, we have that

$$\sum_{i=1}^{\hat{k}} \bar{A}_i \mathscr{R}(B) = \left(0_{(n-k_3)\times k_3}, I_{n-k_3}\right) \hat{Q}^{-1} span\{\widehat{\Phi}_i Bv \mid 1 \le i \le \hat{k}, v \in \mathbb{R}^m\}.$$

Thus, (4.34) is equivalent to the right hand side of (4.33). This ends the proof. □

Next, we introduce symbol matrices.

Definition 4.3 Let α be an l-dimensional vector variable, i.e., $\alpha = (x_1, \ldots, x_l)^*$, where $x_1, \ldots, x_l$ are variables in $\mathbb{R}$. A symbol matrix with respect to α is a matrix whose elements are linear functions of α.

By the definition, when α is an l-dimensional vector variable, an $n' \times m'$ symbol matrix w.r.t. α can be expressed as:

$$\mathscr{A}(\alpha) = \left(\mathscr{A}(\alpha)_{ij}\right)_{i=1,j=1}^{i=n',j=m'},$$
$$\text{with } \mathscr{A}(\alpha)_{ij} = \langle \overrightarrow{a_{ij}}, \alpha \rangle_{\mathbb{R}^l} \text{ for some } \overrightarrow{a_{ij}} \in \mathbb{R}^l.$$

Definition 4.4 Let α be an l-dimensional vector variable and $\mathscr{A}(\alpha)$ be an $n' \times m'$ symbol matrix w.r.t. α. A non-negative integer j is called the rank of $\mathscr{A}(\alpha)$, if (i) any $(j+1)$-order sub-determinant of $\mathscr{A}(\alpha)$ is identically zero; (ii) there is an $\bar{\alpha} \in \mathbb{R}^l$ so that some j-order sub-determinant of $\mathscr{A}(\bar{\alpha})$ is not zero. The rank of $\mathscr{A}(\alpha)$ is denoted by $Rank(\mathscr{A}(\alpha))$.

Given n-dimensional vector variable $\alpha = (x_1, \ldots, x_n)^*$, and $D = (d_{lk}) \in \mathbb{R}^{m' \times n}$ with $m' \in \mathbb{N}$, we define

$$D\alpha \triangleq \left(\sum_{k=1}^{n} d_{1,k} x_k, \ldots, \sum_{k=1}^{n} d_{m'_1,k} x_k \right). \tag{4.35}$$

In what follows, we give n-dimensional vector variables $\alpha_1, \alpha_2, \ldots, \alpha_n$, with $\alpha_i = (x_1^i, \ldots, x_n^i)^*$ for $i = 1, \ldots, n$. We define

$$(\alpha_1, \alpha_2, \ldots, \alpha_j) \triangleq \left(x_1^1, \ldots, x_n^1, x_1^2, \ldots, x_n^2 \ldots, x_1^j, \ldots, x_n^j \right)^* \tag{4.36}$$

for each $1 \le j \le n$. Clearly, it is a jn-dimensional vector variable.

Definition 4.5 Let $\left\{\widehat{\Phi}_i\right\}_{i=1}^{\hat{k}}$ be a family generating $\widehat{\Phi}([0, nT])$ with the affiliated family $\left\{\bar{A}_i\right\}_{i=1}^{\hat{k}}$. (i) A family $\left\{\mathscr{A}_j\right\}_{j=1}^{n}$ of symbol matrices is called the symbol family affiliated to $\left\{\widehat{\Phi}_i\right\}_{i=1}^{\hat{k}}$ if

$$\mathscr{A}_j \triangleq \mathscr{A}_j\left((\alpha_1, \alpha_2, \ldots, \alpha_j)\right) = (\bar{A}_1\alpha_1, \ldots, \bar{A}_{\hat{k}}\alpha_1, \ldots, \bar{A}_1\alpha_j, \ldots, \bar{A}_{\hat{k}}\alpha_j), \quad 1 \le j \le n, \tag{4.37}$$

where $(\alpha_1, \alpha_2, \ldots, \alpha_j)$ is given by (4.36). (ii) When $\left\{\mathscr{A}_j\right\}_{j=1}^{n}$ is the symbol family affiliated to $\left\{\widehat{\Phi}_i\right\}_{i=1}^{\hat{k}}$, $\mathscr{A}_j$ is called the jth symbol matrix affiliated to $\left\{\widehat{\Phi}_i\right\}_{i=1}^{\hat{k}}$.

Clearly, each $\mathscr{A}_j$ is a $(n - k_3) \times (n \times \hat{k})$ symbol matrix w.r.t. the vector variable $(\alpha_1, \alpha_2, \dots, \alpha_j)$.

Definition 4.6 Let $\{\widehat{\Phi}_i\}_{i=1}^{\hat{k}}$ be a family generating $\widehat{\Phi}([0, nT])$, with the affiliated symbol family $\{\mathscr{A}_j\}_{j=1}^{n}$. The symbol subfamily affiliated to $\{\widehat{\Phi}_i\}_{i=1}^{\hat{k}}$ is defined by

$$\mathscr{G} \triangleq \{\mathscr{A}_j \mid Rank\left(\mathscr{A}_j\right) = n - k_3,\ 1 \le j \le n\}. \tag{4.38}$$

Write

$$D\left(\mathscr{G}\right) = \min\{1 \le j \le n \mid \mathscr{A}_j \in \mathscr{G}\}. \tag{4.39}$$

Remark 4.2 We must point out that $\mathscr{G} \neq \varnothing$ and hence, $D(\mathscr{G})$ is well defined. In fact, let $\{\widehat{\Phi}_i\}_{i=1}^{\hat{k}}$ be a family generating $\widehat{\Phi}([0, nT])$ (see Definition 4.1). Let $\{\bar{A}_i\}_{i=1}^{\hat{k}}$ and $\mathscr{A}_n$ be respectively the affiliated family and the nth symbol matrix affiliated to $\{\widehat{\Phi}_i\}_{i=1}^{\hat{k}}$. We claim that $\mathscr{A}_n \in \mathscr{G}$. By Proposition 4.2, we have that $I_n \in \mathscr{C}\mathscr{B}_{A(\cdot)}$. Then, by (4.33) with $B = I_n$, we see that

$$\sum_{i=1}^{\hat{k}} \mathscr{R}(\bar{A}_i) = \mathbb{R}^{n-k_3}.$$

Meanwhile, from (4.37), where $j = n$ and each α_i takes the value e_i, it follows that

$$\mathscr{R}(\mathscr{A}_n(e_1, e_2, \dots, e_n)) = \sum_{i=1}^{\hat{k}} \mathscr{R}(\bar{A}_i).$$

Hence, $Rank\ (\mathscr{A}_n(e_1, e_2, \dots, e_n)) = n - k_3$. Since $\mathscr{A}_n(\alpha_1, \alpha_2, \dots, \alpha_n)$ is an $(n - k_3) \times (n \times \hat{k})$ symbol matrix, we see from Definition 4.4 that $Rank\ (\mathscr{A}_n) = n - k_3$. Then, by (4.38), $\mathscr{A}_n \in \mathscr{G}$.

We now present a way to design a simple control machine in $\mathscr{C}\mathscr{B}_{A(\cdot)}$.

Theorem 4.2 *Let* $\{\widehat{\Phi}_i\}_{i=1}^{\hat{k}}$ *be a family generating* $\widehat{\Phi}([0, nT])$, *with* $\{\bar{A}_i\}_{i=1}^{\hat{k}}$, $\{\mathscr{A}_j\}_{j=1}^{n}$ *and* $\mathscr{G}$ *defined by (4.29), (4.37) and (4.38) respectively. Then,* $M\left(\mathscr{C}\mathscr{B}_{A(\cdot)}\right) = D(\mathscr{G})$. *Moreover, when* $\bar{j} = D(\mathscr{G})$, *any column vectors* $\beta_1, \dots, \beta_{\bar{j}} \in \mathbb{R}^n$, *with*

$$Rank(\bar{A}_1\beta_1, \dots, \bar{A}_{\hat{k}}\beta_1, \dots, \bar{A}_1\beta_{\bar{j}}, \dots, \bar{A}_{\hat{k}}\beta_{\bar{j}}) = n - k_3, \tag{4.40}$$

verifies that

$$B \triangleq \left(\beta_1, \dots, \beta_{\bar{j}}\right) \in \mathscr{C}\mathscr{B}_{A(\cdot)}. \tag{4.41}$$

Proof Let $\bar{j} = D(\mathscr{G})$. By Definition 4.4, there are vectors $\beta_1, \dots, \beta_{\bar{j}}$ in $\mathbb{R}^n$ so that

$$\begin{aligned} &Rank(\mathscr{A}_{\bar{j}}) \\ &= Rank\left((\bar{A}_1\beta_1, \ldots, \bar{A}_{\hat{k}}\beta_1, \bar{A}_1\beta_2, \ldots, \bar{A}_{\hat{k}}\beta_2, \ldots, \bar{A}_1\beta_{\bar{j}}, \ldots, \bar{A}_{\hat{k}}\beta_{\bar{j}})\right) \\ &= n - k_3. \end{aligned}$$

Write $B = \left(\beta_1, \ldots, \beta_{\bar{j}}\right)$. Then we have that

$$\mathbb{R}^{n-k_3} = \mathscr{R}(\mathscr{A}_{\bar{j}}) = \sum_{i=1}^{\hat{k}} \bar{A}_i \mathscr{R}(B).$$

By Proposition 4.4, this yields that $B \in \mathscr{C}\mathscr{B}_{A(\cdot)}$. Hence, $D(\mathscr{G}) = \bar{j} \geq M\left(\mathscr{C}\mathscr{B}_{A(\cdot)}\right)$.

Conversely, write $j_0 \triangleq M\left(\mathscr{C}\mathscr{B}_{A(\cdot)}\right)$. Let $\hat{B} \in \mathscr{C}\mathscr{B}_{A(\cdot)}$ with j_0 columns. Write $\hat{B} = (b_1, \ldots, b_{j_0})$, where b_i is the ith column of $\hat{B}$. Then, by making use of Proposition 4.4 again, we find that

$$span\{\bar{A}_1 b_1, \ldots, \bar{A}_{\hat{k}} b_1, A_1 b_2, \ldots, \bar{A}_{\hat{k}} b_2, \ldots, \bar{A}_1 b_{j_0}, \ldots, \bar{A}_{\hat{k}} b_{j_0}\} = \mathbb{R}^{n-k_3}.$$

This, along with (4.37), yields that $Rank\mathscr{A}_{j_0}((\alpha_1, \ldots, \alpha_{j_0})) = n - k_3$. From this, (4.38) and (4.39), we see that $M\left(\mathscr{C}\mathscr{B}_{A(\cdot)}\right) \triangleq j_0 \geq D(\mathscr{G})$. This ends the proof. □

Remark 4.3 Given a T-periodic $A(\cdot)$, the matrix given by (4.41) is a second kind of simple control machine w.r.t. $A(\cdot)$.

4.3 The Second Kind of Simple Control Machines–Special Case

In this section, we study the second kind of simple control machines in the special case when $A(\cdot)$ is time invariant, i.e., it is an $n \times n$ real matrix. We arbitrarily fix a matrix $A \in \mathbb{R}^{n\times n}$. We will apply Theorem 4.2 to this special case to obtain more precise information on the number $M(\mathscr{C}\mathscr{B}_A)$. Write $\sigma(A) = \{\lambda_1, \ldots, \lambda_r, \mu_1, \bar{\mu}_1, \ldots, \mu_s, \bar{\mu}_s\}$. Here, $\lambda_1, \ldots, \lambda_r$ are distinct real eigenvalues of A and $\mu_1, \bar{\mu}_1, \ldots, \mu_s, \bar{\mu}_s$ are distinct non-real eigenvalues of A. Let m_i, $i = 1, \ldots, r$, and $\hat{m}_j$, $j = 1, \ldots, s$ be respectively the geometric multiplicities of λ_i and μ_j, i.e., $m_i = \dim\mathscr{N}(A - \lambda_i I)$ and $\hat{m}_j = \dim\mathscr{N}(A_c - \mu_j)$ with A_c the complexification of A. Then, m_i is the number of the Jordan blocks associated to λ_i, and $\hat{m}_j$ is the number of the real Jordan blocks associated to $(\mu_j, \bar{\mu}_j)$. The main result for this special case is as follows:

Theorem 4.3 *Let $A \in \mathbb{R}^{n\times n}$. Then $M(\mathscr{C}\mathscr{B}_A) = \max_{\lambda\in\sigma(A)\setminus\mathbb{C}^-} m(\lambda)$, where $m(\lambda)$ is the geometric multiplicity of the eigenvalue λ.*

Before proving this theorem, we introduce some preliminaries. First of all, by the classical matrix theory (see, for instance, Theorem 1 on p. 67 in [40]), there is a non-singular matrix $P \in \mathbb{R}^{n\times n}$ so that

$$\begin{aligned}\Lambda &\triangleq P^{-1}AP \\ &= diag\left\{J_1^{\lambda_1}, \ldots, J_{m_1}^{\lambda_1}, \ldots, J_1^{\lambda_r}, \ldots, J_{m_r}^{\lambda_r}, J_1^{\mu_1}, \ldots, J_{\hat{m}_1}^{\mu_1}, \ldots, J_1^{\mu_s}, \ldots, J_{\hat{m}_s}^{\mu_s}\right\},\end{aligned} \tag{4.42}$$

where $J_l^{\lambda_i} = \begin{pmatrix} \lambda_i & 1 & & \\ & \ddots & \ddots & \\ & & \ddots & 1 \\ & & & \lambda_i \end{pmatrix}$, $J_l^{\mu_i} = \begin{pmatrix} C_i & I_2 & & \\ & \ddots & \ddots & \\ & & \ddots & I_2 \\ & & & C_i \end{pmatrix}$. Here and in what follows, $C_i \triangleq \begin{pmatrix} Re(\mu_i) & Im(\mu_i) \\ -Im(\mu_i) & Re(\mu_i) \end{pmatrix}$, $i = 1, \ldots, \bar{m}$. Write

$$n_{i,l} \triangleq \dim J_l^{\lambda_i}, \quad 1 \le i \le r, 1 \le l \le m_i; \quad 2\hat{n}_{i,l} = \dim J_l^{\mu_i}, \quad 1 \le i \le s, \quad 1 \le l \le \hat{m}_i. \tag{4.43}$$

Let

$$\hat{e}(k) \triangleq (0, \ldots, 0, 1)^* \in \mathbb{R}^k \text{ for each } k \in \mathbb{N}. \tag{4.44}$$

Lemma 4.3 *Let* $\hat{J} = \begin{pmatrix} C & I_2 & & \\ & \ddots & \ddots & \\ & & \ddots & I_2 \\ & & & C \end{pmatrix}$ *be a Jordan block with* $2k$ *order, where* $C = \begin{pmatrix} a & b \\ -b & a \end{pmatrix}$, *with* $a, b \in \mathbb{R}$, $b \neq 0$. *Let* $\hat{e}(2k)$ *be defined by (4.44). Then*

$$span\left\{\hat{e}(2k), \hat{J}\hat{e}(2k), \ldots, \hat{J}^{2k-1}\hat{e}(2k)\right\} = \mathbb{R}^{2k}. \tag{4.45}$$

Proof Write $\{e_1, \ldots, e_{2k}\}$ for the standard basis of $\mathbb{R}^{2k}$. Define a $2k \times 2k$ orthogonal matrix E by $E^* \triangleq (e_1\ e_3\ \ldots e_{2k-1}\ e_2\ e_4\ \ldots e_{2k})$. Then, one can directly check that

$$\widetilde{J} \triangleq E^*\hat{J}E = \begin{pmatrix} A_k & bI_k \\ -bI_k & A_k \end{pmatrix}, \text{ with } A_k = \begin{pmatrix} a & 1 & & \\ & \ddots & \ddots & \\ & & a & 1 \\ & & & a \end{pmatrix} \text{ a } k \times k \text{ matrix,}$$

and that $\hat{e}(2k) = E\hat{e}(2k)$. Let $\hat{I}_{2k} = \begin{pmatrix} \frac{1}{2}I_k & \frac{1}{2}I_k \\ \frac{i}{2}I_k & -\frac{i}{2}I_k \end{pmatrix}$. Hence, we have that

$$\tilde{J} = \hat{I}_{2k}\begin{pmatrix} J_k & \\ & \bar{J}_k \end{pmatrix}\left(\hat{I}_{2k}\right)^{-1}, \quad \text{with } J_k = \begin{pmatrix} a+ib & 1 & & \\ & \ddots & \ddots & \\ & & \ddots & 1 \\ & & & a+ib \end{pmatrix} \in \mathbb{R}^{k\times k}.$$

Thus, one can directly check that

$$\begin{aligned} &\left|\hat{e}(2k), \hat{J}\hat{e}(2k), \ldots, \hat{J}^{2k-1}\hat{e}(2k)\right| \\ &= |E|\left|\hat{I}_{2k}\right|\left|\begin{pmatrix} \hat{e}(k) \\ \hat{e}(k) \end{pmatrix}, \begin{pmatrix} J_k\hat{e}(k) \\ \bar{J}_k\hat{e}(k) \end{pmatrix}, \ldots, \begin{pmatrix} J_k^{2k-1}\hat{e}(k) \\ \bar{J}_k^{2k-1}\hat{e}(k) \end{pmatrix}\right|, \end{aligned}$$

where $|S|$ denotes to the determinant of a square matrix S. From this, (4.45) follows at once, provided that the following matrix is invertible:

$$H \triangleq \left(\begin{pmatrix} \hat{e}(k) \\ \hat{e}(k) \end{pmatrix}, \begin{pmatrix} J_k\hat{e}(k) \\ \bar{J}_k\hat{e}(k) \end{pmatrix}, \ldots, \begin{pmatrix} J_k^{2k-1}\hat{e}(k) \\ \bar{J}_k^{2k-1}\hat{e}(k) \end{pmatrix}\right).$$

To prove that H is invertible, we let $(a_0, a_1, \ldots, a_{2k-1})^* \in \mathbb{C}^{2k}$ verify that

$$H(a_0, a_1, \cdots, a_{2k-1})^* = 0_{2k\times 1}. \tag{4.46}$$

Define a polynomial:

$$f(\lambda) \triangleq a_{2k-1}\lambda^{2k-1} + a_{2k-2}\lambda^{2k-2} + \cdots + a_1\lambda + a_0.$$

From (4.46), we see that

$$f(J_k)\hat{e}(k) = f(\bar{J}_k)\hat{e}(k) = 0. \tag{4.47}$$

Hence, there are polynomials g_1, g_2, h_1, h_2 (with complex coefficients) so that

$$f(\lambda) = g_1(\lambda)(\lambda - a - ib)^k + h_1(\lambda), \; f(\lambda) = g_2(\lambda)(\lambda - a + ib)^k + h_2(\lambda),$$

where $\partial(h_1) < k, \; \partial(h_2) < k$ (Here, $\partial(h_i)$ denotes the degree of h_i). Because

$$(J_k - (a+ib)I_k)^k = 0 = \left(\bar{J}_k - (a-ib)I_k\right)^k,$$

it follows from (4.47) that $h_1(J_k)\hat{e}(k) = 0$ and $h_2(\bar{J}_k)\hat{e}(k) = 0$. This, along with the invertibility of matrices:

$$\left(\hat{e}(k), J_k\hat{e}(k), \ldots, J_k^{k-1}\hat{e}(k)\right) \quad \text{and} \quad \left(\hat{e}(k), \bar{J}_k\hat{e}(k), \ldots, \bar{J}_k^{k-1}\hat{e}(k)\right),$$

implies that $h_1(\lambda) \equiv 0$ and $h_2(\lambda) \equiv 0$. Thus, f is a multiple of $(\lambda - a - ib)^k(\lambda - a - ib)^k$. This indicates that $f(\lambda) \equiv 0$, since $\partial(f) < 2k$. Therefore, $(a_0, a_1, \ldots, a_{2k-1})^* = 0$ and matrix H is non-singular. This ends the proof. $\square$

Proposition 4.5 *Let $A_1 \in \mathbb{R}^{p\times p}$, $A_2 \in \mathbb{R}^{q\times q}$, $B_1 \in \mathbb{R}^{p\times r}$, $B_2 \in \mathbb{R}^{q\times r}$. Suppose that the following systems are controllable:*

$$\dot{x}(t) = A_1x(t) + B_1u(t),\ t \geq 0 \ \text{and}\ \dot{y}(t) = A_2y(t) + B_2u(t),\ t \geq 0.$$

Assume that $\sigma(A_1)\bigcap\sigma(A_2) = \varnothing$. Then the following system is also controllable:

$$\frac{d}{dt}\begin{pmatrix} x \\ y \end{pmatrix}(t) = \begin{pmatrix} A_1 & \\ & A_2 \end{pmatrix}\begin{pmatrix} x \\ y \end{pmatrix}(t) + \begin{pmatrix} B_1 \\ B_2 \end{pmatrix}u(t). \tag{4.48}$$

Proof Let $(x_0, y_0) \in \mathbb{R}^p \times \mathbb{R}^q$ verify that

$$(B_1^*, B_2^*)\begin{pmatrix} e^{A_1^*t} & \\ & e^{A_2^*t} \end{pmatrix}\begin{pmatrix} x_0 \\ y_0 \end{pmatrix} = 0,\ \ t \geq 0. \tag{4.49}$$

We aim to show that $(x_0, y_0) = 0$. When this is done, the controllability of (4.48) follows from the classical O.D.E control theory (see, for instance, Theorem 1.7 on p. 112 in [59]). Clearly, (4.49) is equivalent to

$$B_1^*e^{A_1^*t}x_0 + B_2^*e^{A_2^*t}y_0 = 0 \ \text{ for all }\ t \geq 0.$$

By differentiating the above equality times, we obtain that

$$B_1^*e^{A_1^*t}(A_1^*)^kx_0 + B_2^*e^{A_2^*t}(A_2^*)^ky_0 = 0,\ \ k = 0, 1, 2, \ldots. \tag{4.50}$$

Let $f_1(\cdot)$ and $f_2(\cdot)$ be respectively the characteristic polynomials of A_1 and A_2. Because $\sigma(A_1)\bigcap\sigma(A_2) = \varnothing$, $f_1(\cdot)$ and $f_2(\cdot)$ are coprime. Thus there are polynomials $g_1(\cdot)$ and $g_2(\cdot)$ so that

$$g_1(\cdot)f_1(\cdot) + g_2(\cdot)f_2(\cdot) \equiv 1. \tag{4.51}$$

It follows from (4.50) that

$$B_1^*e^{A_1^*t}(g_1 \times f_1)(A_1)x_0 + B_2^*e^{A_2^*t}(g_1 \times f_1)(A_2)y_0 = 0. \tag{4.52}$$

On the other hand, by the Hamilton-Cayley Theorem, we have that $f_1(A_1) = 0$ and $f_2(A_2) = 0$. These, as well as (4.50), (4.51) and (4.52), yields that

$$\begin{aligned} B_2^*e^{A_2^*t}y_0 &= B_2^*e^{A_2^*t}Iy_0 = B_2^*e^{A_2^*t}(g_1 \times f_1 + g_2 \times f_2)(A_2)y_0 \\ &= B_2^*e^{A_2^*t}(g_1 \times f_1)(A_2)y_0 = -B_1^*e^{A_1^*t}(g_1 \times f_1)(A_1)x_0 = 0. \end{aligned}$$

Since $[A_2, B_2]$ is controllable, the above implies that $y_0 = 0$. Similarly, we can verify that $x_0 = 0$. This completes the proof. □

Since the Kalman rank condition for $[A, B]$ is equivalent to the controllability of $[A, B]$, we have following consequence of Proposition 4.5:

Corollary 4.2 *Suppose that*

$$Rank(B_1, A_1 B_1, \ldots, A_1^{p-1} B_1) = p, \qquad Rank(B_2, A_2 B_2, \ldots, A_2^{q-1} B_2) = q.$$

Assume that $\sigma(A_1) \bigcap \sigma(A_2) = \emptyset$. *Then*

$$Rank\left(\begin{pmatrix} B_1 \\ B_2 \end{pmatrix}, \begin{pmatrix} A_1 & \\ & A_2 \end{pmatrix}\begin{pmatrix} B_1 \\ B_2 \end{pmatrix}, \ldots, \begin{pmatrix} A_1 & \\ & A_2 \end{pmatrix}^{p+q-1}\begin{pmatrix} B_1 \\ B_2 \end{pmatrix}\right) = p + q.$$

We are now on the position to prove Theorem 4.3.

Proof of Theorem 4.3. We first prove the equality in Theorem 4.3 for the case where $\sigma(A) \bigcap \mathbb{C}^- = \emptyset$. In this case, $k_3 = 0$ and $\hat{Q} = I$, where k_3 and $\hat{Q}$ are given by (4.12) and (4.13) respectively. By the Hamilton-Cayley theorem,

$$span\left\{e^{-At}, t \in [0, nT]\right\} = span\left\{I, A, A^2, \ldots, A^{n-1}\right\}.$$

Hence, $\{I, A, \ldots, A^{n-1}\}$ is a family generating $\widehat{\Phi}([0, nT])$ with $\widehat{\Phi}(t) = e^{-tA}$. By Definitions 4.2, 4.5, the family and the symbol family affiliated to $\left\{I, A, A^2, \ldots, A^{n-1}\right\}$ are respectively $\{A^{i-1}\}_{i=1}^n$ and $\{\mathscr{A}_j\}_{j=1}^n$ with

$$\mathscr{A}_j \triangleq \mathscr{A}_j((\alpha_1, \ldots, a_j)) = (\alpha_1, \ldots, A^{n-1}\alpha_1, \ldots, \alpha_j, \ldots, A^{n-1}\alpha_j) \tag{4.53}$$

for any $1 \leq j \leq n$. Let

$$\hat{\mathscr{A}_j} \triangleq \hat{\mathscr{A}_j}((\alpha_1, \ldots, \alpha_j)) \triangleq (\alpha_1, \ldots, \Lambda^{n-1}\alpha_1, \alpha_j, \ldots, \Lambda^{n-1}\alpha_j), \ 1 \leq j \leq n, \tag{4.54}$$

where Λ is given by (4.42). Clearly, $\hat{\mathscr{A}_j}$ is an $n \times nj$ symbol matrix w.r.t. $(\alpha_1, \ldots, \alpha_j)$, and

$$Rank\big(\mathscr{A}_j((a_1, \ldots, a_j))\big) = Rank\big(\hat{\mathscr{A}_j}(\alpha_1, \ldots, a_j))\big). \tag{4.55}$$

Write

$$\bar{m} = \max\{m_1, \ldots, m_r, \hat{m}_1, \ldots, \hat{m}_s\}. \tag{4.56}$$

By Theorem 4.2 and Definition 4.6 (where $k_3 = 0$), the equality in Theorem 4.3 is equivalent to the following two properties:

$$Rank\big(\hat{\mathscr{A}_j}((\alpha_1, \ldots, \alpha_j))\big) < n, \quad \text{when } \ j < \bar{m}, \tag{4.57}$$

and

$$Rank\left(\hat{\mathscr{A}}_{\bar{m}}((\alpha_1, \ldots, \alpha_{\bar{m}}))\right) = n. \tag{4.58}$$

We organize the proof of (4.57) and (4.58) by the following two steps:
Step 1: The proof of (4.57)

Suppose that $j < \bar{m}$. By Definition 4.4 and (4.55), to prove (4.57), it suffices to show that each n-order sub-determinant of $\hat{\mathscr{A}}_j((\alpha_1, \ldots, a_j))$ is 0. Seeking for a contradiction, we suppose that it did not stand. Then, there would be vectors $\bar{\alpha}_1, \bar{\alpha}_2, \ldots, \bar{\alpha}_j \in \mathbb{R}^n$ so that one of n-order sub-determinant of $\hat{\mathscr{A}}_j\left((\hat{\alpha}_1, \hat{\alpha}_2, \ldots, \hat{\alpha}_j)\right)$ is not zero. Since the matrix $\hat{\mathscr{A}}_j\left((\bar{\alpha}_1, \bar{\alpha}_2, \ldots, \bar{\alpha}_j)\right)$ has exactly n rows, any group of distinct row vectors in $\hat{\mathscr{A}}_j\left((\bar{\alpha}_1, \bar{\alpha}_2, \ldots, \bar{\alpha}_j)\right)$ is linearly independent. Without loss of generality, we can assume that either $\bar{m} = m_1$ or $\bar{m} = \hat{m}_1$. We will deduce a contradiction from each one of the above-mentioned two cases.

In the first case where $\bar{m} = m_1$, we write

$$n_l \triangleq \sum_{i=1}^{l} n_{1,i}, \quad l = 1, 2, \ldots, \bar{m}, \tag{4.59}$$

where $n_{1,i}$ is defined by (4.43). Let

$$\hat{\alpha}_i \triangleq (e_{n_1} \ldots e_{n_{\bar{m}}})^* \bar{\alpha}_i \in \mathbb{R}^{\bar{m}}, \qquad i = 1, \ldots, j, \tag{4.60}$$

where and throughout the proof, $\{e_1, \ldots, e_n\}$ stands for the standard basis of $\mathbb{R}^n$ with each e_i a column vector. Notice that if we write $\bar{\alpha}_i = (x_1^i, x_2^i, \ldots, x_n^i)^*$, with $x_l^i \in \mathbb{R}$, then $\hat{\alpha}_i = (x_{n_1}^i, x_{n_2}^i, \ldots, x_{n_{\bar{m}}}^i)^*$. From (4.42) and (4.60), one has that when $1 \leq i \leq j$ and $k \in \mathbb{N}$,

$$\left(e_{n_1}, e_{n_2}, \ldots, e_{n_{\bar{m}}}\right)^* \Lambda^k \bar{\alpha}_i = \begin{pmatrix} e_{n_1}^* \\ e_{n_2}^* \\ \ldots \\ e_{n_{\bar{m}}}^* \end{pmatrix} \Lambda^k \bar{\alpha}_i = \lambda_1^k \begin{pmatrix} e_{n_1}^* \\ e_{n_2}^* \\ \ldots \\ e_{n_{\bar{m}}}^* \end{pmatrix} \bar{\alpha}_i = \lambda_1^k \hat{\alpha}_i.$$

This, along with (4.54), indicates that

$$\begin{aligned} &\left(e_{n_1}, e_{n_2}, \ldots, e_{n_{\bar{m}}}\right)^* \hat{\mathscr{A}}_j(\bar{\alpha}_1, \bar{\alpha}_2, \ldots, \bar{\alpha}_j) \\ &= \left(e_{n_1}, \ldots, e_{n_{\bar{m}}}\right)^* (\bar{\alpha}_1, \ldots, \Lambda^{n-1}\bar{\alpha}_1, \ldots, \bar{\alpha}_j, \ldots, \Lambda^{n-1}\bar{\alpha}_j) \\ &= (\hat{\alpha}_1, \lambda_1\hat{\alpha}_1, \ldots, \lambda_1^{n-1}\hat{\alpha}_1, \ldots, \hat{\alpha}_j, \lambda_1\hat{\alpha}_j, \ldots, \lambda_1^{n-1}\hat{\alpha}_j). \end{aligned} \tag{4.61}$$

Clearly, any maximal independent group of column vectors in the matrix on the right side of (4.61) has at most j vectors. Thus, the rank of the matrix on the right side of (4.61) is less than or equals to $j (< \bar{m})$.

On the other hand, the matrix on the left side of (4.61) consists of $\bar{m}$ row vectors which are exactly n_1th, n_2th, ... and $n_{\bar{m}}$th rows in $\hat{\mathscr{A}}_j(\bar{\alpha}_1, \bar{\alpha}_2, \ldots, \bar{\alpha}_j)$. Since the rank of the matrix on the right hand side of (4.61) is less than $\bar{m}$, these row vectors are linearly dependent, which contradicts with the fact that any group of distinct row vectors in $\hat{\mathscr{A}_j}\big((\bar{\alpha}_1, \bar{\alpha}_2, \ldots, \bar{\alpha}_j)\big)$ is linearly independent.

In the second case where $\bar{m} = \hat{m}_1$, we let

$$\tilde{n} = \dim\left(diag\left\{J_1^{\lambda_1}, \ldots, J_{m_1}^{\lambda_1}, \ldots, J_1^{\lambda_r}, \ldots, J_{m_r}^{\lambda_r}\right\}\right) \tag{4.62}$$

and

$$\hat{n}_l = \tilde{n} + \sum_{i=1}^{l} 2\hat{n}_{1,i}, \quad l = 1, 2, \ldots, \bar{m}, \tag{4.63}$$

where $\hat{n}_{1,i}$ is defined by (4.43). By (4.42), one can easily check that

$$\begin{pmatrix} e^*_{\hat{n}_l - 1} \\ e^*_{\hat{n}_l} \end{pmatrix} \Lambda^k = C_1^k \begin{pmatrix} e^*_{\hat{n}_l - 1} \\ e^*_{\hat{n}_l} \end{pmatrix}, \quad l = 1, \ldots, \bar{m}, \ k \in \mathbb{N}.$$

For each $\bar{\alpha}_i$, with $i \in \{1, \cdots, j\}$, we define a vector

$$\widetilde{\alpha}_i \triangleq (e_{\hat{n}_1 - 1} \ \ e_{\hat{n}_1} \ \ e_{\hat{n}_2 - 1} \ \ e_{\hat{n}_2} \ldots e_{\hat{n}_{\bar{m}} - 1} \ \ e_{\hat{n}_{\bar{m}}})^* \bar{\alpha}_i \in \mathbb{R}^{2\bar{m}}. \tag{4.64}$$

Notice that, if we denote $\bar{\alpha}_i = (x_1^i, \ldots, x_n^i)^*$, with each $x_k^i \in \mathbb{R}$, then

$$\widetilde{\alpha}_i = (x^i_{\hat{n}_1 - 1}, x^i_{\hat{n}_1}, \ldots, x^i_{\hat{n}_{\bar{m}} - 1}, x^i_{\hat{n}_{\bar{m}}}) \triangleq (\widetilde{\alpha}^*_{i1}, \widetilde{\alpha}^*_{i2}, \ldots, \widetilde{\alpha}^*_{i\bar{m}}), \ i = 1, \ldots j, \tag{4.65}$$

where $\widetilde{\alpha}^*_{ik} = (x^i_{\hat{n}_k - 1}, x^i_{\hat{n}_k})^*, \ \ k = 1, \ldots, \bar{m}$. By (4.42), (4.54), (4.64) and (4.65), one can directly verify that

$$\begin{aligned} &\left(e_{\hat{n}_1 - 1} \ \ e_{\hat{n}_1} \ \ e_{\hat{n}_2 - 1} \ \ e_{\hat{n}_2}, \ldots, e_{\hat{n}_{\bar{m}} - 1} \ \ e_{\hat{n}_{\bar{m}}}\right)^* \hat{\mathscr{A}_j}(\bar{\alpha}_1, \bar{\alpha}_2, \ldots, \bar{\alpha}_j) \\ &= \begin{pmatrix} \widetilde{\alpha}_{11} & C_1\widetilde{\alpha}_{11} & \cdots & C_1^{n-1}\widetilde{\alpha}_{11} & \cdots & \widetilde{\alpha}_{j1} & C_1\widetilde{\alpha}_{j1} & \cdots & C_1^{n-1}\widetilde{\alpha}_{j1} \\ \widetilde{\alpha}_{12} & C_1\widetilde{\alpha}_{12} & \cdots & C_1^{n-1}\widetilde{\alpha}_{12} & \cdots & \widetilde{\alpha}_{j2} & C_1\widetilde{\alpha}_{j2} & \cdots & C_1^{n-1}\widetilde{\alpha}_{j2} \\ \cdots & \cdots & \cdots & \cdots & \cdots & \cdots & \cdots & \cdots & \cdots \\ \widetilde{\alpha}_{1\bar{m}} & C_1\widetilde{\alpha}_{1\bar{m}} & \cdots & C_1^{n-1}\widetilde{\alpha}_{1\bar{m}} & \cdots & \widetilde{\alpha}_{j\bar{m}} & C_1\widetilde{\alpha}_{j\bar{m}} & \cdots & C_1^{n-1}\widetilde{\alpha}_{j\bar{m}} \end{pmatrix}, \end{aligned} \tag{4.66}$$

Because C_1 is a 2×2 matrix, by the Hamilton Cayley theorem, each C_1^k, with $k \in \mathbb{N}$, is a linear combination of I_2 and C_1. Thus, any maximal independent group of column vectors in the matrix on the right hand side of (4.66) has at most $2j$ vectors. Therefore, the rank of the matrix on the right hand side of (4.66) is less than or equals to $2j\,(<2\bar{m})$. On the other hand, the matrix on the left hand side of (4.66) consists of $2\bar{m}$ row vectors which are exactly $(\hat{n}_1 - 1)$th, $\hat{n}_1$th, ..., $(\hat{n}_{\bar{m}} - 1)$th and $\hat{n}_{\bar{m}}$th rows of $\hat{\mathscr{A}_j}(\bar{\alpha}_1, \bar{\alpha}_2, \ldots, \bar{\alpha}_j)$. Since the rank of the matrix on the right hand side of (4.66) is less than $2\bar{m}$, these $2\bar{m}$ row vectors are linearly dependent, which contradicts to the fact that any group of distinct row vectors in $\hat{\mathscr{A}_j}\big((\bar{\alpha}_1, \bar{\alpha}_2, \ldots, \bar{\alpha}_j)\big)$ is linearly independent. So we have proved (4.57).

Step 2: The proof of (4.58)
By Definition 4.4, it suffices to structure vectors $\xi_1, \ldots, \xi_{\bar{m}}$ in $\mathbb{R}^n$ so that

$$Rank\left(\left(\hat{\mathscr{A}}_{\bar{m}}((\xi_1, \xi_2, \ldots, \xi_{\bar{m}}))\right)\right) = n. \tag{4.67}$$

For this purpose, we let $\hat{e}(k)$, with $k \in \mathbb{N}$, be given by (4.44). Then, we construct, for each $k \in \{1, \ldots, \bar{m}\}$, a vector ξ_k in $\mathbb{R}^n$ by setting

$$\xi_k^* \triangleq \Big((f_{1,1}^k)^* \ \ldots \ (f_{1,m_1}^k)^* \ \ldots \ (f_{r,1}^k)^* \ldots (f_{r,m_r}^*)^* \\ (g_{1,1}^k)^* \ldots (g_{1,\hat{m}_1}^k)^* \ldots (g_{s,1}^k)^k \ldots (g_{s,\hat{m}_s}^k)^* \Big),$$

where $f_{i,l}^k \in \mathbb{R}^{n_{i,l}}$, with $1 \leq i \leq r$ and $1 \leq l \leq m_i$, is defined in the following manner:

$$f_{i,l}^k = \hat{e}(n_{n_{i,l}}), \text{ when } l = k; \qquad f_{i,l}^k = 0_{n_{i,l}\times 1}, \text{ when } l \neq k.$$

and where $g_{i,l}^k \in \mathbb{R}^{2\hat{n}_{i,l}}$ with $1 \leq i \leq s$ and $1 \leq l \leq \hat{m}_i$, is defined in the following manner:

$$g_{i,l}^k = \hat{e}(2\hat{n}_{n_{i,l}}), \text{ when } l = k; \qquad g_{i,l}^k = 0_{2n_{i,l}\times 1}, \text{ when } l \neq k.$$

By (4.42) and Lemma 4.3, we find that

$$span\left\{\hat{e}(n_{i,l}), J_l^{\lambda_i}\hat{e}(n_{i,l}), \ldots, (J_l^{\lambda_i})^{n_{i,l}-1}\hat{e}(n_{i,l})\right\} = \mathbb{R}^{n_{i,l}}, \tag{4.68}$$

when $1 \leq i \leq r$, $1 \leq l \leq m_i$;

$$span\left\{\hat{e}(2\hat{n}_{i,l}), J_l^{\mu_i}\hat{e}(2\hat{n}_{i,l}), \ldots, (J_l^{\mu_i})^{2\hat{n}_{i,l}-1}\hat{e}(2\hat{n}_{i,l})\right\} = \mathbb{R}^{2\hat{n}_{i,l}}, \tag{4.69}$$

when $1 \leq s$, $1 \leq l \leq \hat{m}_i$. Write $\mathscr{N}_k$, with $1 \leq k \leq \bar{m}$, for the set of all such $k' \in \{1, 2, \ldots, n\}$ that the k'th row of Λ contains one of rows in one of blocks $J_k^{\lambda_1}, \ldots, J_k^{\lambda_r}, \hat{J}_k^{\mu_1}, \ldots, \hat{J}_k^{\mu_s}$ (where $J_k^{\lambda_i} = \varnothing$ or $\hat{J}_k^{\mu_j} = \varnothing$ if $m_i < k$ or $\hat{m}_j < k$). From (4.42), the definition of ξ_k^*, (4.68), (4.69) and Corollary 4.2, we see that

$$span\left\{\xi_k, \ \Lambda\xi_k, \ \ldots, \ \Lambda^{n-1}\xi_k\right\} = span\{e_l, \ l \in \mathscr{N}_k\}. \tag{4.70}$$

From the definition of $\mathscr{N}_k$, we find that

$$\bigcup_{k=1}^{\bar{m}} \mathscr{N}_k = \{1, 2, \ldots, n\} \text{ and } \mathscr{N}_i \bigcap \mathscr{N}_j = \varnothing, \text{ when } i \neq j. \tag{4.71}$$

Thus, by the definition of $\hat{\mathscr{A}}_{\bar{m}}((\alpha_1, \alpha_2, \ldots, \alpha_{\bar{m}}))$ (see (4.54)), (4.70) and (4.71), we obtain (4.67). So (4.58) stands. In summary, we have proved the equality in Theorem 4.3 for the case where $\sigma(A) \bigcap \mathbb{C}^- = \varnothing$.

The remaining is to prove the equality in Theorem 4.3 for the case that $\sigma(A) \bigcap \mathbb{C}^- \neq \varnothing$. In this case, we can apply Lemma 1.2 to decouple the system $x' = Ax$ into two subsystems: $x_1' = A_1 x_1$ and $x_2' = A_2 x_2$, with $\sigma(A_1) \subset \mathbb{C}^-$ and $\sigma(A_2) \cap \mathbb{C}^- = \varnothing$ (where the first one is k_3-order, while the second one is $(n-k_3)$-order). If $(n-k_3) \times k$-order matrix B_2 belongs to $\mathscr{CB}_{A_2}$, then we have that $\begin{pmatrix} 0_{k_3 \times k} \\ B_2 \end{pmatrix} \in \mathscr{CB}_A$, which leads to

$$M\left(\mathscr{CB}_A\right) \leq M\left(\mathscr{CB}_{A_2}\right). \tag{4.72}$$

Meanwhile we notice $\mathscr{P} = diag\left\{e^{A_1 T}, e^{A_2 T}\right\}$. In (4.12), select $Q = diag\left\{I_{k_3}, I_{n-k_3}\right\}$. Write $\widehat{\Phi}_i \triangleq diag\left\{A_1^{i-1}, A_2^{i-1}\right\}$ with $i = 1, \ldots, n$. Then by the Hamilton-Cayley theorem, $\left\{\widehat{\Phi}_i\right\}_{i=1}^n$ is a family generating $\widehat{\Phi}([0, nT])$ (see (4.28)). Then the family affiliated to $\left\{\widehat{\Phi}_i\right\}_{i=1}^n$ is $\{\bar{A}_i\}_{i=1}^n$ with $\bar{A}_i \triangleq \left(0_{(n-k_3)\times k_3}, A_2^{i-1}\right)$. For any $B \in \mathscr{CB}_A$, assume that k is the column number of B. Let $B_2 \triangleq \left(0_{(n-k_3)\times k_3}, I_{n-k_3}\right) B \in \mathbb{R}^{(n-k_3)\times k}$. By Proposition 4.4, one can directly check that

$$\begin{aligned}
B \in \mathscr{CB}_A &\Leftrightarrow \sum_{i=1}^n \bar{A}_i \mathscr{R}(B) = \mathbb{R}^{n-k_3} \\
&\Leftrightarrow \sum_{i=1}^n A_2^{i-1}\left(0_{(n-k_3)\times k_3}, I_{n-k_3}\right) \mathscr{R}(B) = \mathbb{R}^{n-k_3} \\
&\Leftrightarrow \sum_{i=1}^n A_2^{i-1} \mathscr{R}(B_2) = \mathbb{R}^{n-k_3} \\
&\Leftrightarrow B_2 \in \mathscr{CB}_{A_2}.
\end{aligned}$$

So $M\left(\mathscr{CB}_{A_2}\right) \leq M\left(\mathscr{CB}_A\right)$. This, together with (4.72), indicates that $M\left(\mathscr{CB}_A\right) = M\left(\mathscr{CB}_{A_2}\right)$. This ends the proof. □

Remark 4.4 In the proof above, we indeed provide a way to structure a matrix B with the column number $M\left(\mathscr{CB}_A\right)$.

We end this chapter with two examples. In the first example, we will apply Theorem 4.2 to find $B \in \mathscr{CB}_{A(\cdot)}$ for a given $A(\cdot)$.

Example 4.1 Let

$$A(t) \triangleq \begin{pmatrix} 0 & -\pi\{t\} \\ \pi\{t\} & 0 \end{pmatrix}, \quad t \geq 0,$$

where $\{t\}$ denotes the decimal part of t. It is 1-periodic. We will apply Theorem 4.2 to find $B \in \mathscr{CB}_{A(\cdot)}$ with the minimum number of columns. First of all, one can directly check that the fundamental solution, associated with this $A(\cdot)$, is given by

$$\widehat{\Phi}(t) = \begin{cases} \begin{pmatrix} \cos(\pi t^2/2) & -\sin(\pi t^2/2) \\ \sin(\pi t^2/2) & \cos(\pi t^2/2) \end{pmatrix} & \text{when } t \in [0, 1]; \\ \begin{pmatrix} -\sin(\pi(t-1)^2/2) & -\cos(\pi(t-1)^2/2) \\ \cos(\pi(t-1)^2/2) & -\sin(\pi(t-1)^2/2) \end{pmatrix} & \text{when } t \in (1, 2], \end{cases}$$

and

$$\widehat{\Phi}(t+2) = -\widehat{\Phi}(t) \text{ when } t \in [0, 2] \quad \text{and} \quad \widehat{\Phi}(t+4) = \widehat{\Phi}(t) \text{ when } t \in [0, +\infty).$$

Hence, $\widehat{\Phi}(\cdot)$ is 4-periodic. Because $\mathscr{P} = \widehat{\Phi}(1)$, it follows that

$$\sigma(\mathscr{P}) = \{i, -i\} \quad \text{and} \quad \sigma(\mathscr{P}) \cap \mathbb{B} = \emptyset.$$

From these and (4.12), we can take $\hat{Q} = I_2$. Then (4.13), we see that $k_3 = 0$. Meanwhile, by a direct calculation, we find that

$$span\{\widehat{\Phi}^{-1}(t) \mid t \in [0, 2]\} = \mathbb{R}^{2\times 2}.$$

This, along with (4.21), yields that $\bar{k} = 4$. Now we let

$$\widehat{\Phi}_1 \triangleq \widehat{\Phi}(0)^{-1} = \begin{pmatrix} 1 & 0 \\ 0 & 1 \end{pmatrix}, \quad \widehat{\Phi}_2 \triangleq \widehat{\Phi}(\sqrt{1/2})^{-1} = \begin{pmatrix} \sqrt{2}/2 & \sqrt{2}/2 \\ -\sqrt{2}/2 & \sqrt{2}/2 \end{pmatrix},$$
$$\widehat{\Phi}_3 \triangleq \widehat{\Phi}(\sqrt{2/3})^{-1} = \begin{pmatrix} 1/2 & \sqrt{3}/2 \\ -\sqrt{3}/2 & 1/2 \end{pmatrix}, \quad \widehat{\Phi}_4 \triangleq \widehat{\Phi}(1)^{-1} = \begin{pmatrix} 0 & 1 \\ -1 & 0 \end{pmatrix}.$$

Then one can easily check that $\{\widehat{\Phi}_1, \widehat{\Phi}_2, \widehat{\Phi}_3, \widehat{\Phi}_4\}$ is a family generating $\widehat{\Phi}([0, 2])$ with $\hat{k} = \bar{k} = 4$ (see Definition 4.1). Let

$$\bar{A}_i = \widehat{\Phi}_i, \quad i = 1, 2, 3, 4.$$

Since $\hat{Q} = I_2$ and $k_3 = 0$, one can easily check that $\{\bar{A}_j\}_{j=1}^4$ is the family affiliated to $\{\widehat{\Phi}_1, \widehat{\Phi}_2, \widehat{\Phi}_3, \widehat{\Phi}_4\}$ (see Definition 4.2).

Next, we take two 2-dimensional vector variables (see Definition 4.3):

$$\alpha_i = \begin{pmatrix} x_{i1} \\ x_{i2} \end{pmatrix}, \quad i = 1, 2.$$

Let

$$\mathscr{A}_1 = \begin{pmatrix} x_{11} & \frac{\sqrt{2}}{2}(x_{11} + x_{12}) & \frac{1}{2}x_{11} + \frac{\sqrt{3}}{2}x_{12} & x_{12} \\ x_{12} & \frac{\sqrt{2}}{2}(x_{12} - x_{11}) & \frac{1}{2}x_{12} - \frac{\sqrt{3}}{2}x_{11} & -x_{11} \end{pmatrix},$$

$$\mathscr{A}_{22} = \begin{pmatrix} x_{21} & \frac{\sqrt{2}}{2}(x_{21} + x_{22}) & \frac{1}{2}x_{21} + \frac{\sqrt{3}}{2}x_{22} & x_{22} \\ x_{22} & \frac{\sqrt{2}}{2}(x_{22} - x_{21}) & \frac{1}{2}x_{22} - \frac{\sqrt{3}}{2}x_{21} & -x_{21} \end{pmatrix},$$

and $\mathscr{A}_2 = (\mathscr{A}_1, \mathscr{A}_{22})$. Then one can easily check that $\{\mathscr{A}_1, \mathscr{A}_2\}$ is a symbol family affiliated to $\{\widehat{\Phi}_1, \widehat{\Phi}_2, \widehat{\Phi}_3, \widehat{\Phi}_4\}$ (see Definition 4.5). By a direct calculation, we see that (see Definition 4.4)

$$Rank(\mathscr{A}_1) = Rank(\mathscr{A}_2) = 2,$$

and

$$\mathscr{G} = \{\mathscr{A}_1,\ \mathscr{A}_2\} \quad \text{with } D(\mathscr{G}) = 1.$$

Finally, by applying Theorem 4.2, we find that

$$M(\mathscr{C}\mathscr{B}_{A(\cdot)}) = D(\mathscr{G}) = 1.$$

By a direct verification, we see that for any nonzero column vector $B \in \mathbb{R}^2$,

$$Rank(\bar{A}_1 B, \bar{A}_1 B, \bar{A}_3 B, \bar{A}_4 B) = 2.$$

By making use of Theorem 4.2 again, we obtain that for any column vector $B \in \mathbb{R}^2 \backslash \{0\}$, $B \in \mathscr{C}\mathscr{B}_{A(\cdot)}$ with the minimum number of columns.

The next example concerns designs of a kind of simple control machines in infinitely dimensional cases.

Example 4.2 Consider the following 1-periodic heat equation

$$\begin{cases} \partial_t y(x,t) - \Delta y(x,t) - 9 \cdot \{t\} \cdot y(x,t) = \chi_{(0,\pi/2)} u(x,t) \text{ in } (0,\pi) \times \mathbb{R}^+ \\ y(0,t) = y(\pi,t) = 0 \qquad\qquad\qquad\qquad\qquad\qquad \text{in } \mathbb{R}^+, \end{cases} \tag{4.73}$$

where $u \in L^2(\mathbb{R}^+; U)$, with U a subspace of $L^2((0,\pi))$ and $\{t\}$ denotes the decimal part of t. First of all, $\{\lambda_k\}_{k=1}^{\infty}$, with $\lambda_k \triangleq k^2$, is the family of all eigenvalues for $-\Delta$ with the Dirichlet boundary condition and $\{\xi_k\}_{k=1}^{\infty}$, with $\xi_k = \sin kx$, is the family of the corresponding eigenfunctions, which serves as a basis of $L^2((0,\pi))$. By direct computations, we find that for each $t \in \mathbb{R}^+$,

$$\mathscr{P}(t)z = \sum_{k=1}^{\infty} e^{-k^2+9/2} a_k \xi_k, \quad \text{when } \ z = \sum_{k=1}^{\infty} a_k \xi_k \text{with} \{a_k\}_{k=1}^{\infty} \in l^2.$$

Let $H_1(t)$ and $H_2(t)$ be the corresponding subspaces given by (1.21) in Proposition 1.4. Let n_0 be given by (1.17). Then we find from Proposition 1.4 that

$$H_1(t) \equiv H_1 = span\{\sin x,\ \ \sin 2x\} \ \text{ for all } \ t \geq 0$$

$$H_2(t) \equiv H_2 = span\{\sin 3x,\ \ \sin 4x,\ \ldots\} \ \text{ for all } \ t \geq 0$$

and

$$n_0 = 2.$$

Here, we used the fact that $e^{-k^2+9/2} \geq 1$ only when $k = 1, 2$. When $U = L^2((0, \pi))$, it follows from Corollary 2.1 in Sect. 2.3.2 that there is a subspace $\hat{Z}$ of $L^2(\Omega)$ with $\dim Z \leq 2$ so that (4.73) is LPFS with respect to $\hat{Z}$.

In what follows, we arbitrarily fix a subspace $U \subset L^2((0, \pi))$ so that (4.73) is LPFS. Let $\mathscr{U}^{FS}$ be given by (2.1) corresponding to this U. We aim to find a subspace Z from $\mathscr{U}^{FS}$ so that it has the minimal dimension. This leads to the minimization problem:

$$N(U) \triangleq \min\Big\{\dim Z \;\Big|\; Z \in \mathscr{U}^{FS}\Big\}. \tag{4.74}$$

For the infinitely dimensional system (4.73), the way of control acting in the system depends not only on the control operator, but also on the choice of Z. The choice of this subspace can be interpreted as a way of designs of control machines. Hence, a solution of (4.74) can be treated as a kind of simple control machine.

Now we are going to find a solution to (4.74). Since $n_0 = 2$. it follows by Theorem 2.2 that $N(U) \leq 2$. Because (4.73), with the null control, is unstable, $N(U)$ can take two possible values 1 or 2. Write

$$U_1 \triangleq \Big\{v \in U \;\Big|\; \int_0^{\pi/2} v(x) \sin x \mathrm{d}x \neq 0\Big\},$$
$$U_2 \triangleq \Big\{v \in U \;\Big|\; \int_0^{\pi/2} v(x) \sin 2x \mathrm{d}x \neq 0\Big\}.$$

We claim that

$$U_1 \neq \varnothing \quad \text{and} \quad U_2 \neq \varnothing. \tag{4.75}$$

Suppose, by contradiction, that (4.75) were not true. Then, we would have that either $U_1 = \varnothing$ or $U_2 = \varnothing$. In the case that $U_1 = \varnothing$, by letting $\hat{H}_2 = H_2 + span\{\xi_2\}$, we find that $\chi_{(0,\pi/2)} u \in L^2(\mathbb{R}^+; \hat{H}_2)$ for any $u \in L^2(\mathbb{R}^+; U)$. From this, we can apply Theorem 2.1 to get that (4.73) is not LPFS with respect to U, which leads to a contradiction, since we assumed that (4.73) is LPFS with respect to U. In the case that $U_2 = \varnothing$, we can similarly get a contradiction. Hence, (4.75) is true.

There are only two cases about U_1 and U_2: $U_1 \cap U_2 \neq \varnothing$ and $U_1 \cap U_2 = \varnothing$. In the first case that $U_1 \cap U_2 \neq \varnothing$, we can easily check, by using Theorem 2.1, that $N(U) = 1$ and that for any $v \in U_1 \cap U_2$, $Z \triangleq span\{v\} \in \mathscr{U}^{FS}$ is a solution to (4.74). In the second case that $U_1 \cap U_2 = \varnothing$, we can easily check that $N(U) = 2$ and that for any $v_1 \in U_1$ and $v_2 \in U_2$, $Z \triangleq span\{v_1, v_2\} \in \mathscr{U}^{FS}$ is a solution to (4.74).

Miscellaneous Notes

By our understanding, the procedure to stabilize periodically a system: $\dot{y}(t) = A(t)y(t)$ (where $A(\cdot) \in L^\infty(\mathbb{R}^+; \mathbb{R}^{n\times n})$ is T-periodic) is as follows: One first builds up a T-periodic $B(\cdot) \in L^\infty(\mathbb{R}^+; \mathbb{R}^{n\times m})$ so that $[A(\cdot), B(\cdot)]$ is T-periodically stablizable, and then design a periodic (such as T-periodic or nT-period) $K(\cdot) \in$

$L^\infty(\mathbb{R}^+; \mathbb{R}^{m\times n})$ so that $A(\cdot) + B(\cdot)K(\cdot)$ is exponentially stable. We call the aforementioned $B(\cdot)$ as a control machine and the corresponding $K(\cdot)$ as a feedback law. Control machines could be treated as control equipments which belong to the category of hardware, while feedback laws could be treated as control programs which belong to the category of software. Thus, it is interesting to answer to following question: How to design a simple T-periodic $B(\cdot)$ for a given T-periodic $A(\cdot)$ so that $[A(\cdot), B(\cdot)]$ is T-periodically stabilizable? A matrix $\hat{B}$, with $M\left(\mathscr{C}\mathscr{B}_{A(\cdot)}\right)$ columns, in $\mathscr{C}\mathscr{B}_{A(\cdot)}$ (see (4.3) and (4.4)) could be one of the simplest ones. From this point of view, we give an answer for the above question, with the help of Theorem 3.3. In the control theory of PDEs, the studies on the optimal location and the optimal shape of controllers (see, for instance, [76–79]) are indeed quite closed to what we did above, i.e., to ask for a *simple* control machine. Besides, for infinitely dimensional systems, the way of control acting in systems is not fully prescribed by the control operator when controls take value in a subspace of a control space. The choice of this subspace can be interpreted as a way of design of control machine. Our Example 4.2 interprets it.

The material in this chapter is taken from [92].

References

1. Aeyels, D.: Stabilization of a class of nonlinear systems by a smooth feedback control. Syst. Control Lett. **5**, 289–294 (1985)
2. Arnold, V.I.: Ordinary Differential Equations. Springer, Berlin (2003)
3. Bacciotti, A., Boieri, P.: Linear stabilizability of planar nonlinear systems. Math. Control Signals Systems **3**, 183–193 (1990)
4. Balakrishnan, A.V.: Boundary control of parabolic equations: L-Q-R theory. In: Proceedings of Fifth Internat. Summer School. Central Inst. Math. Mech. Acad. Sci. GDR, Berlin (1977)
5. Barbu, V.: Feedback stabilization of Navier-Stokes equations. ESAIM Control Optim. Calc. Var. **9**, 197–206 (2003)
6. Barbu, V.: Stabilization of Navier-Stokes Flows. Springer, London (2011)
7. Barbu, V., Wang, G.: Feedback stabilization of periodic solutions to nonlinear parabolic-like evolution systems. Indiana Univ. Math. J. **54**, 1521–1546 (2005)
8. Bellman, R., Glicksberg, I., Gross, O.: Some Aspects of the Mathematical Theory of Control Processes. Rand Co., Santa Monica (1958)
9. Bensoussan, A., Da Prato, G., Delfour, M.C., Mitter, S.K.: Representation and Control of Infinite Dimensional Systems. Systems & Control: Foundations & Applications, vol. I. Birkhäuser Boston Inc., Boston (1992)
10. Bensoussan, A., Da Prato, G., Delfour, M.C., Mitter, S.K.: Representation and Control of Infinite Dimensional Systems. Systems & Control: Foundations & Applications, vol. II. Birkhäuser Boston Inc., Boston (1993)
11. Bensoussan, A., Da Prato, G., Delfour, M.C., Mitter, S.K.: Representation and Control of Infinite Dimensional Systems. Systems & Control: Foundations & Applications, 2nd edn. Birkhäuser Boston Inc., Boston (2007)
12. Bhattacharyya, S.P.: Robust Stabilization Against Structured Perturbations. Lecture Notes in Control and Information Sciences, vol. 99. Springer, Berlin (1987)
13. Bisognin, E., Bisognin, V., Pereira, J.M.: Exponential stabilization of periodic solutions of a system of KdV equations. Port. Math. **66**, 191–210 (2009)
14. Bochner, S.: Integration von Funktionen, deren Werte die Elemente eines Vectorraumes sind. Fundamenta Mathematicae **20**, 262–276 (1933)
15. Boothby, W.M., Marino, R.: Feedback stabilization of planar nonlinear systems. Systems Control Lett. **12**, 87–92 (1989)
16. Brockett, R.W.: Asymptotic stability and feedback stabilization. In: Differential Geometric Control Theory (Houghton, Mich., 1982). Progress in Mathematics, vol. 27, pp. 181–191. Birkhäuser Boston, Boston (1983)
17. Brockett, R.W.: A stabilization problem. Open Problems in Mathematical Systems and Control Theory. Communications and Control Engineering, pp. 75–78. Springer, London (1999)

G. Wang and Y. Xu, *Periodic Feedback Stabilization for Linear Periodic Evolution Equations*, SpringerBriefs in Mathematics,
DOI 10.1007/978-3-319-49238-4

18. Brunovský, P.: Controllability and linear closed-loop controls in linear periodic systems. J. Diff. Eqs. **6**, 296–313 (1969)
19. Chen, G.: Energy decay estimates and exact boundary value controllability for the wave equation in a bounded domain. J. Math. Pures Appl. **58**, 249–273 (1979)
20. Chen, G.: Control and stabilization for the wave equation in a bounded domain. SIAM J. Control Optim. **17**, 66–81 (1979)
21. Chen, G., Russell, D.L.: A mathematical model for linear elastic systems with structural damping. Q. Appl. Math. **39**, 433–454 (1982)
22. Clarke, F.H., Ledyaev, Y.S., Sontag, E.D., Subbotin, A.I.: Asymptotic controllability implies feedback stabilization. IEEE Trans. Automat. Control. **42**, 1394–1407 (1997)
23. Conrad, F.: Stabilization of beams by pointwise feedback control. SIAM J. Control Optim. **28**, 423–437 (1990)
24. Coron, J.M.: Control and Nonlinearity. American Mathematical Society, vol. 136. Mathematical Surveys and Monographs, Providence (2007)
25. Coron, J., Praly, L., Teel, A.: Feedback stabilization of nonlinear systems: sufficient conditions and Lyapunov and input-output techniques. In: Trends in Control (Rome, 1995), pp. 293-348. Springer, Berlin (1995)
26. Curtain, R.F., Pritchard, A.J.: the infinite-dimensional Riccati equation for systems defined by evolution operators. SIAM J. Control Optim. **14**, 951–983 (1976)
27. Curtain, R.F., Pritchard, A.J.: An abstract theory for unbounded control action for distributed parameter systems. SIAM J. Control. **15**, 566–611 (1977)
28. Curtain, R.F., Pritchard, A.J.: Infinite Dimensional Linear Systems Theory. Lecture Notes in Control and Information Sciences, vol. 8. Springer, Berlin (1978)
29. Da Prato, G.: Linear quadratic control theory for infinite dimensional systems. In: Mathematical Control Theory, Part 1, 2 (Trieste, 2001). ICTP Lecture Notes, vol. VIII, pp. 59–105 (electronic). Abdus Salam International Centre for Theoretical Physics, Trieste (2002)
30. Da Prato, G., Ichikawa, A.: Quadratic control for linear time-varying systems. SIAM J. Control Optim. **28**, 359–381 (1990)
31. Da Prato, G., Lunardi, A.: Floquent exponents and Stabilitizability in Time-periodic parobolic systems. Appl. Math. Optim. **22**, 91–113 (1990)
32. Diestel, J., Uhl Jr., J.J.: Vector Measures. Mathematical Surveys, vol. 15. AMS, Providence (1977)
33. Fabre, C., Puel, J.P., Zuazua, E.: Approximate controllability of the semilinear heat equation. Proc. Royal Soc. Edinburgh. **125A**, 31–61 (1995)
34. Fleming, W.H.: Future Directions in Control Theory, a Mathematical Perspective. SIAM, Philadelphia (1988)
35. Floquet, G.: Sur les équations différentielles linéaires à coefficients périodiques. Ann. Sci. École Norm. Sup. **12**, 47–88 (1883). (in French)
36. Friedman, B.M.: Economic Stabilization Policy: Methods in Optimization. Studies in Mathematical and Managerial Economics, vol. 15. North-Holland Publishing Co., Amsterdam (1975). American Elsevier Publishing Co., New York
37. Hansen, S.W.: Exponential energy decay in a linear thermoelastic rod. J. Math. Anal. Appl. **167**, 429–442 (1992)
38. Henry, D.: Geometric Theory of Semilinear Parabolic Equations. Lecture Notes in Mathematics, vol. 840. Springer, Berlin (1981)
39. Hermes, H.: On the synthesis of a stabilizing feedback control via Lie algebraic methods. SIAM J. Control Optim. **18**, 352–361 (1980)
40. Hirsch, M.W., Smale, S.: Differential Equations, Dynamical Systems, and Linear Algebra. Pure and Applied Mathematics, vol. 60. Academic Press, New York (1974). (A subsidiary of Harcourt Brace Jovanovich, Publishers)
41. Huang, F.L.: Characteristic condition for exponential stability of linear dynamical systems in Hilbert spaces. Ann. Diff. Eqs. **1**, 43–56 (1985)
42. Jurdjevic, V., Quinn, J.P.: Controllability and stability. J. Diff. Eqs. **28**, 381–389 (1978)

43. Kalman, R.E.: Contributions to the theory of optimal control. Bol. Soc. Math. Mexicana **5**, 102–119 (1960)
44. Kano, H., Nishimura, T.: Periodic solution of matrix Riccati equations with detectability and stabilizability. Internat. J. Control **29**, 471–487 (1979)
45. Kano, H., Nishimura, T.: Controllability, stabilizability, and matrix Riccati equations for periodic systems. IEEE Trans. Automat. Control **30**, 1129–1131 (1985)
46. Kato, T.: Perturbation Theory for Linear Operators, Die Grundlehren der mathematischen Wissenschaften, vol. 132. Springer, New York (1966)
47. Kawski, M.: Stabilization of nonlinear systems in the plane. Systems Control Lett. **12**, 169–175 (1989)
48. Komornik, V.: Exact Controllability and Stabilization: The Multiplier Method. RAM: Research in Applied Mathematics. Masson, Paris (1994). Wiley, Chichester
49. Komornik, V., Russell, D.L., Zhang, B.Y.: Stabilisation de l'équation de Korteweg-de Vries (in French). C. R. Acad. Sci. Paris Sér. I Math. **312**, 841-843 (1991)
50. Krstić, M., Deng, H.: Stabilization of Nonlinear Uncertain Systems. Communications and Control Engineering Series. Springer, London (1998)
51. Kurepa, S.: On the quadratic functional. Acad. Serbe Sci. Publ. Inst. Math. **13**, 57–72 (1961)
52. Kwan, C.C., Wang, K.N.: Sur la stabilisation de la vibration elastique. Scientia Sinica. **17**, 446–467 (1974)
53. Lagnese, J.E.: Boundary Stabilization of Thin Plates. SIAM Studies in Applied Mathematics, vol. 10. SIAM, Philadelphia (1989)
54. Lasiecka, I., Triggiani, R.: Differential and Algebraic Riccati Equations with Applications to Boundary/Point Control Problems: Continuous Theory and Approximation Theory. Lecture Notes in Control and Information Sciences, vol. 164. Springer, Berlin (1991)
55. Lasiecka, I., Triggiani, R.: Control Theory for Partial Differential Equations: Continuous and Approximation Theories, I, Abstract Parabolic Systems. Encyclopedia of Mathematics and its Applications, vol. 74. Cambridge University Press, Cambridge (2000)
56. Laurent, C., Rosier, L., Zhang, B.: Control and stabilization of the Korteweg-de Vries equation on a periodic domain. Comm. Partial Diff. Eqs. **35**, 707–744 (2010)
57. Lee, K.K., Arapostathis, A.: Remarks on smooth feedback stabilization of nonlinear systems. Systems Control Lett. **10**, 41–44 (1988)
58. Leonov, G.A.: The Brockett stabilization problem (in Russian). Avtomat. i Telemekh, 190–193 (2001); translation in. Autom. Remote Control **62**, 847–849 (2001)
59. Li, X., Yong, J.: Optimal Control Theory for Infinite-Dimensional Systems. Systems & Control: Foundations & Applications. Birkhäuser Boston Inc., Boston (1995)
60. Lin, F.H.: A uniqueness theorem for parabolic equations. Comm. Pure Appl. Math. **43**, 127–136 (1990)
61. Lions, J.L.: Sur le contrôle optimal de systèmes décrits par des équations aux dérivées partielles linéaires, Remarques générales; Équations elliptiques; Équations d'évolution. C. R. Acad. Sci. Paris, Ser. A-B. **263**, 661–663; 713–715; 776–779 (1966)
62. Lions, J.L.: Optimal Control of Systems Governed by Partial Differential Equations. Springer, New York (1971)
63. Lions, J.L.: Exact controllability, stabilization and perturbations for distributed systems. SIAM Review. **30**, 1–68 (1988)
64. Liu, K.S.: Energy decay problems in the design of a point stabilizer for coupled string vibrating systems. SIAM J. Control Optim. **26**, 1348–1356 (1988)
65. Liu, K.S., Huang, F.L., Chen, G.: Exponential stability analysis of a long chain of coupled vibrating strings with dissipative linkage. SIAM J. Appl. Math. **49**, 1694–1707 (1989)
66. Lukes, D.L., Russell, D.L.: The quadratic criterion for distributed systems. SIAM J. Control. **7**, 101–121 (1969)
67. Lunardi, A.: Stabilizability of time-periodic parabolic equations. SIAM J. Control Optim. **29**, 810–828 (1991)

68. Lunardi, A.: Neumann boundary stabilization of structurally damped time periodic wave and plate equations. Differential Equations with Applications in Biology, Physics, and Engineering (Leibnitz 1989). Lecture Notes in Pure and Applied Mathematics, vol. 133, pp. 241–257. Dekker, New York (1991)
69. Luo, Z., Guo, B., Morgul, O.: Stability and Stabilization of Infinite Dimensional Systems with Applications. Communications and Control Engineering Series. Springer, London (1999)
70. Lyapunov, A.M.: The general problem of the stability of motion, Translated from Edouard Davaux's French translation (1907) of the 1892 Russian original and edited by A. T. Fuller, with an introduction and preface by Fuller, a biography of Lyapunov by V. I. Smirnov, and a bibliography of Lyapunov's works compiled by J. F. Barrett, Lyapunov centenary issue, Reprint of. Int. J. Control **55**, 521–790 (1992)
71. Malkin, I.G.: The Stability Theory of Motion. Nauk Press, Moscow (1966). (in Russian)
72. Micu, S., Zuazua, E.: Stabilization and periodic solutions of a hybrid system arising in the control of noise. Control of partial differential equations and applications (Laredo 1994). Lecture Notes in Pure and Applied Mathematics, vol. 174, pp. 219–230. Dekker, New York (1996)
73. Penrose, R.: A generalized inverse for matrices. Proc. Cambridge Philos. Soc. **51**, 406–413 (1955)
74. Phung, K.D., Wang, G.: Quantitative unique continuation for the semilinear heat equation in a convex domain. J. Funct. Anal. **259**, 1230–1247 (2010)
75. Phung, K.D., Wang, G.: An observability estimate for parabolic equations from a measurable set in time and its applications. J. Eur. Math. Soc. **15**, 681–703 (2013)
76. Privat, Y., Trélat, E., Zuazua, E.: Optimal observation of the one-dimensional wave equation. J. Fourier Anal. Appl. **19**, 514–544 (2013)
77. Privat, Y., Trélat, E., Zuazua, E.: Optimal location of controllers for the one-dimensional wave equation. Ann. Inst. H. Poincaré Anal. Non Linéaire **30**, 1097–1126 (2013)
78. Privat, Y., Trélat, E., Zuazua, E.: Optimal observability of the multi-dimensional wave and Schrödinger equations in quantum ergodic domains. J. Eur. Math. Soc. **18**, 1043–1111 (2016)
79. Privat, Y., Trélat, E., Zuazua, E.: Optimal shape and location of sensors for parabolic equations with random initial data. Arch. Rational Mech. Anal. **216**, 921–981 (2015)
80. Raymond, J.: Feedback boundary stabilization of the two-dimensional Navier-Stokes equations. SIAM J. Control Optim. **45**, 790–828 (2006)
81. Russell, D.L.: Controllability and stabilizability theory for linear partial differential equations: recent progress and open problems. SIAM Rev. **20**, 639–739 (1978)
82. Russell, D.L., Zhang, B.: Stabilization of the Korteweg-de Vries equation on a periodic domain. Control and optimal design of distributed parameter systems (Minneapolis, MN, 1992). IMA Mathematics and its Applications, vol. 70, pp. 195–211. Springer, New York (1995)
83. Saberi, A., Stoorvogel, A., Sannuti, P.: Internal and External Stabilization of Linear Systems with Constraints. Systems & Control: Foundations & Applications. Birkhäuser/Springer, New York (2012)
84. van der Schaft, A.J.: Stabilization of Hamiltonian systems. Nonlinear Anal. **10**, 1021–1035 (1986)
85. Slemrod, M.: Stabilization of bilinear control systems with applications to nonconservative problems in elasticity. SIAM J. Control Optimization. **16**, 131–141 (1978)
86. Sontag, E.D.: Mathematical Control Theory: Deterministic Finite-dimensional Systems. Texts in Applied Mathematics, vol. 6, 2nd edn. Springer, New York (1998)
87. Sontag, E.D.: Feedback stabilization of nonlinear systems. Robust control of linear systems and nonlinear control (Amsterdam 1989). Progress in Systems and Control Theory, vol. 4, pp. 61–81. Birthäuser Boston, Boston (1990)
88. Steeb, W.H., Hardy, Y.: Matrix calculus and Kronecker product. A Practical Approach to Linear and Multilinear Algebra, 2nd edn. World Scientific Publishing Co., Pte. Ltd., Hackensack (2011)
89. Sun, S.H.: On spectrum distribution of completely controllable linear systems. SIAM J. Control Optim. **19**, 730–743 (1981)

90. Tarn, T.J., Zavgren Jr., J.R., Zeng, X.: Stabilization of infinite-dimensional systems with periodic feedback gains and sampled output. Automatica J. IFAC. **24**, 95–99 (1988)
91. Tsinias, J.: Sufficient Lyapunov-like conditions for stabilization. Math. Control Signals Systems **2**, 343–357 (1989)
92. Wang, G., Xu, Y.: Periodic stabilization for linear time-periodic ordinary differential equations. ESAIM Control Optim. Calc. Var. **20**, 269–314 (2014)
93. Wang, G., Xu, Y.: Equivalent conditions on periodic feedback stabilization for linear periodic evolution equations. J. Funct. Anal. **266**, 5126–5173 (2014)
94. Yamé, J.J., Hanus, R.: On stabilization and spectrum assignment in periodically time-varying continuous systems. IEEE Trans. Automat Contr. **46**, 979–983 (2001)
95. Yong, J., Zheng, S.: Feedback stabilization and optimal control for Cahn-Hilliard equation. Nonlinear Anal. **17**, 431–444 (1991)
96. Yong, J., Zheng, S.: Feedback stabilization for the phase field equations. Appl. Anal. **42**, 59–68 (1991)
97. Yosida, K.: Functional Analysis. Grundlehren der Mathematischen Wissenschaften [Fundamental Principles of Mathematical Sciences], vol. 123, 6th edn. Springer, Berlin (1980)
98. Zabczyk, J.: Some comments on stabilizability. Appl. Math. Optim. **19**, 1–9 (1989)
99. Zhang, B., Zhao, X.: Control and stabilization of the Kawahara equation on a periodic domain. Commun. Inf. Syst. **12**, 77–95 (2012)

Index

G. Wang and Y. Xu, *Periodic Feedback Stabilization for Linear Periodic Evolution Equations*, SpringerBriefs in Mathematics,
DOI 10.1007/978-3-319-49238-4

Printed in the United States
By Bookmasters